D0619841

Mathematics and Sports

The Dolciani Mathematical Expositions

NUMBER FORTY-THREE

Mathematics and Sports

Edited by

Joseph A. Gallian
University of Minnesota Duluth

Published and Distributed by
The Mathematical Association of America

DOLCIANI MATHEMATICAL EXPOSITIONS

Committee on Books
Paul Zorn, *Chair*

Dolciani Mathematical Expositions Editorial Board
Underwood Dudley, *Editor*
Jeremy S. Case
Rosalie A. Dance
Tevian Dray
Patricia B. Humphrey
Virginia E. Knight
Michael J. McAsey
Mark A. Peterson
Jonathan Rogness
Thomas Q. Sibley

The DOLCIANI MATHEMATICAL EXPOSITIONS series of the Mathematical Association of America was established through a generous gift to the Association from Mary P. Dolciani, Professor of Mathematics at Hunter College of the City University of New York. In making the gift, Professor Dolciani, herself an exceptionally talented and successful expositor of mathematics, had the purpose of furthering the ideal of excellence in mathematical exposition.

The Association, for its part, was delighted to accept the gracious gesture initiating the revolving fund for this series from one who has served the Association with distinction, both as a member of the Committee on Publications and as a member of the Board of Governors. It was with genuine pleasure that the Board chose to name the series in her honor.

The books in the series are selected for their lucid expository style and stimulating mathematical content. Typically, they contain an ample supply of exercises, many with accompanying solutions. They are intended to be sufficiently elementary for the undergraduate and even the mathematically inclined high-school student to understand and enjoy, but also to be interesting and sometimes challenging to the more advanced mathematician.

MAA Service Center
P.O. Box 91112
Washington, DC 20090-1112
1-800-331-1MAA   FAX: 1-301-206-9789

Preface

Each year the Joint Policy Board for Mathematics (JPBM) sponsors Mathematics Awareness Month to increase public understanding of and appreciation for mathematics. This is achieved through a web page, posters, resource materials, and theme essays. With its abundance of data, great variety, numerous strategies, and widespread popularity, sports is an ideal venue to demonstrate the illuminating power of mathematics to a larger audience. This book is an eclectic compendium of the essays solicited for the 2010 Mathematics Awareness Month on the theme Mathematics and Sports.

In keeping with the goal of promoting mathematics awareness to a broad audience, all of the articles are accessible to college level mathematics students and many are accessible to the general public.

The book is divided into sections by the kind of sports. The section on football includes articles on ranking college football teams; counting the number of defensive formations an NFL team can deploy; and evaluating a method for reducing the advantage of the winner of a coin flip in an NFL overtime game.

The section on track and field examines the ultimate limit on how fast a human can run 100 meters; ranking college track and field conferences; ways to determine the winning team in a cross-country race; the effects of wind and altitude in a 400 meter race; and modeling the biomechanics of running and walking.

The section on baseball has essays on ways to measure the performance of baseball players; deciding if humidifying baseballs reduces the number of home runs; and the probability of batting streaks.

The section on golf has articles on strategies in golf; modeling a golf swing; breaking down Tiger Woods game; and modeling Tiger Woods career.

For basketball there are articles about strategies in basketball at the end of the game; modeling jump shots; and ranking college basketball teams.

For tennis there is an article about strategies in tennis and an essay about using tennis to teach mathematics.

The last section features articles about how to schedule a tournament; the aerodynamics of a free kick in soccer; and tire design for NASCAR cars.

I am grateful to Woody Dudley for meticulously reading every article and providing valuable advice for improving the exposition.

Joe Gallian
University of Minnesota Duluth

Contents

Part I

Baseball

CHAPTER 1

Sabermetrics: The Past, the Present, and the Future

Jim Albert

Abstract

This article provides an overview of sabermetrics, the science of learning about base-ball through objective evidence. Statistics and baseball have always had a strong kinship, as many famous players are known by their famous statistical accomplish-ments, such as Joe Dimaggio's 56-game hitting streak and Ted Williams' .406 batting average in the 1941 baseball season. We give an overview of how one measures per-formance in batting, pitching, and fielding. In baseball, the traditional measures are batting average, slugging percentage, and on-base percentage, but modern measures such as OPS (on-base percentage plus slugging percentage) are better in predicting the number of runs a team will score in a game. Pitching is a harder aspect of perfor-mance to measure, since traditional measures such as winning percentage and earned run average are confounded by the abilities of the pitcher's teammates. Modern mea-sures of pitching such as DIPS (defense independent pitching statistics) are helpful in isolating the contributions of a pitcher that do not involve his teammates. It is also challenging to measure the quality of a player's fielding ability, since the standard measure of fielding, the fielding percentage, is not helpful in understanding the range of a player in moving towards a batted ball. New measures of fielding have been developed that are useful in measuring a player's fielding range. Major League Base-ball is measuring the game in new ways, and sabermetrics is using this new data to find better measures of player performance. The trajectory and speed for all pitches are currently being measured using the Pitch F/X system, and this article demon-strates how this new data can be used to measure the quality of a pitcher's fastball. New data measuring the location of all batted balls and the fielders will lead to fur-ther improvements in measures of player performance which will help teams better understand player values.

1.1 Introduction

Baseball fans have always had a love for statistics. Even when professional baseball began in 1876, counts of the basic statistics such as hits, doubles, triples, home runs, walks, strikeouts, and runs were recorded. Pitchers and hitters have always been ranked with respect to measures such as the batting average, the number of home runs, and the average runs allowed. Currently, one of the most prestigious achievements for hitting is the Triple Crown, when a player simultaneously has the highest batting average, slugging percentage, and number of home runs. (The last player to obtain the Triple Crown was Carl Yastrzemski in 1967.)

Sabermetrics is the science of learning about baseball through objective evidence. Sabermetrics poses questions such as "How many home runs will Albert Pujols hit next year?", "Is it easier to hit home runs in particular ballparks?", "Are particular players especially good in clutch situations?" and collects and summarizes relevant data to answer them.

One basic problem in sabermetrics is evaluating the performance of batters, pitchers, and fielders. We give an overview of the traditional measures for evaluating players, describe some of the current measures that have been developed, and look forward to new evaluation methods based on new types of data collection.

1.2 Measuring Batting

The traditional measure of batting performance is the batting average AVG that is computed by dividing the number of hits H by the number of at-bats AB:

$$AVG = \frac{H}{AB}.$$

The player with the highest batting average is recognized as the batting champion. But this is a flawed measure of batting performance for several reasons. First, the batting average ignores other ways for a player to reach base such as getting a walk or being hit by a pitch. An alternative statistic that measures a player's ability to reach on-base is the on-base percentage (OBP) that divides the total number of on-base events (hits, walks, and hit-by-pitches) by the number of plate appearances:

$$OBP = \frac{H + BB + HBP}{AB + BB + HBP + SF}.$$

Another flaw of the batting average is that it gives all hits the same value and does not distinguish between singles, doubles, triples, and home runs. An alternative measure that distinguishes the different hit values is the slugging percentage (SLG) that computes the average number of bases reached for each at-bat:

$$SLG = \frac{1B + 2 \times 2B + 3 \times 3B + 4 \times HR}{AB},$$

where $1B$, $2B$, $3B$, and HR denote respectively the count of singles, doubles, triples and home runs.

There are two important skills for a batter. He wishes to get on base and, when there are already runners on base, he wishes to advance them home. The on-base percentage measures the effectiveness of a batter in reaching base and the slugging percentage is useful in measuring the batter's skill in advancing runners. One way of measuring the combined skill of the batter to get on base and advance runners adds the on-base percentage to the slugging percentage, creating the modern statistic OPS:

$$OPS = OBP + SLG.$$

Four ways of measuring hitting have been presented: the traditional batting average AVG, the on-base percentage OBP, the slugging percentage SLG, and the combined statistic OPS. Which is the best measure? To answer this question, we have to realize that the goal of batting is to score runs, and runs are scored by teams and not individuals. So it is necessary to look at team hitting data to evaluate the worth of different measures.

We can measure the ability of a team to score by R/G, the average number of runs scored per game. For each team in the 2008 season, we collect R/G, the team batting average, the team on-base percentage, the team slugging percentage, and the team OPS. Figure 1.1 displays a scatterplot of batting average against R/G. As expected, there is a positive trend in this graph — teams with higher batting averages tend to score more runs. But the points are widely scattered which indicates that AVG is not a good predictor of runs scored. Only 46 percent of the total variation in runs scored can be explained by batting average, so 54 percent of the variability in runs scored is due to other differences between the thirty teams.

Figure 1.2 shows a scatterplot of a good hitting measure OPS and R/G. Here we see a strong positive trend in the graph indicating that this measure that combines the on-base percentage and the slugging percentage is a good predictor of runs scored. In fact, 89 percent of the total variation in runs scored can be explained by the differences in OPS.

Major League Baseball and the baseball media can be slow to adopt new measures of performance. But OPS is one measure that has become popular

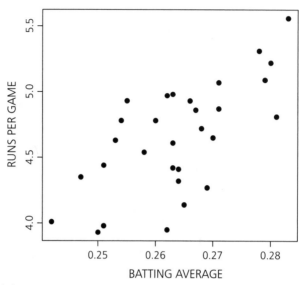

Figure 1.1. Scatterplot of batting average and runs scored per game for all teams in the 2008 baseball season. There is a relatively weak positive association pattern in the scatterplot, indicating that batting average is a weak predictor of runs scored.

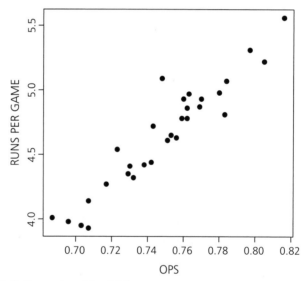

Figure 1.2. Scatterplot of the *OPS* measure and runs scored per game for all teams in the 2008 baseball season. There is a strong positive association pattern in the graph, indicating that *OPS* is a good predictor of runs scored.

Table 1.1.

Name	AVG	OBP	SLG	$OPS = AVG + SLG$
Ichiro Suzuki	0.351	0.396	0.431	0.827
Raul Ibanez	0.291	0.351	0.480	0.831

and is useful in comparing players. For example, Ichiro Suzuki hit for a batting average of .351 in the 2007 season. It appears that he was a much better hitter than his teammate Raul Ibanez, who only had a .291 batting average. But if we compare their OPS (Table 1.1), we see that although Suzuki was more successful in getting on base, Ibanez was more successful in advancing runners. The two players had similar values of OPS in the 2007 season, which means they created the same number of runs for their teams. An OPS of 1.000 is a standard of great performance. Only nine players in baseball history have had a career OPS exceeding 1.000. Babe Ruth is arguably the greatest hitter in baseball history with the highest career OPS value of 1.164.

1.3 Measuring Pitching

How does Major League Baseball currently evaluate pitchers? A traditional measure of pitching is the pitcher's win/loss percentage

$$WIN\% = \frac{W}{W + L},$$

where W and L are respectively the number of team wins and team losses credited to the pitcher. Another traditional measure is the earned run average (ERA), which is the average number of earned runs allowed by the pitcher in 9 innings:

$$ERA = 9 \times \frac{Earned\ Runs}{Innings\ Pitched}.$$

Both measures have some problems. A pitcher can have a high winning percentage not because he is a great player, but because his team tends to score many runs when he is pitching. The earned run average seems like a good measure — after all, a pitcher's objective is to prevent runs scored and the ERA gives the number of runs allowed per game. But there are two problems with an ERA. Teams prevent runs by good pitching and good fielding and the ERA reflects the combined effort of the pitcher and his teammates. The ERA does not separate the ability of the pitcher from the ability of the fielders. Another problem with the ERA is there is a chance element in preventing runs. A pitcher cannot control the path of a ball that is placed in-play, a ball hit in the fair territory of the baseball field. So a pitcher may have a high *ERA* not because he is a poor pitcher, but because many balls happened by chance to fall as base hits.

One way of demonstrating that a particular hitting or pitching statistic is a good measure of ability is to compare its value in one season with its value of the statistic in the following season. If there is a strong relationship between the two statistics, this particular measure is a good measure of the player's ability and we can expect it to make a good prediction of its value for the following season.

Is the earned run average a good measure of a pitcher's ability? For all the pitchers who started at least 25 games in both the 2007 and 2008 baseball seasons, Figure 1.3 shows a scatterplot of the 2007 *ERA* and the 2008 *ERA*. Although there is a positive trend in the scatterplot, only 9 percent in the variability in the 2008 *ERA* values can be explained by the variability in the 2007 *ERA* values. This is a remarkable finding. It indicates that a pitcher's *ERA* is controlled largely by factors such as the team's fielding and lucky balls that fall for base hits that are outside the control of the pitcher.

Sabermetricians have searched for alternative measures of pitching performance that are independent of the contributions made by fielders. The pitcher's ability to strike out batters does not depend on his teammates, and

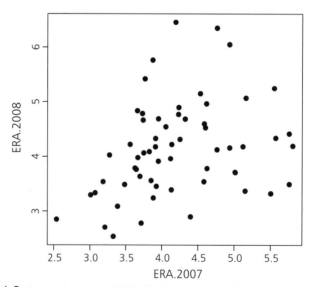

Figure 1.3. Scatterplot of the 2007 *ERA* and the 2008 *ERA* for all pitchers who started at least 25 games in both seasons. There is a weak association pattern in the graph, indicating that the *ERA* is a weak measure of a pitcher's ability.

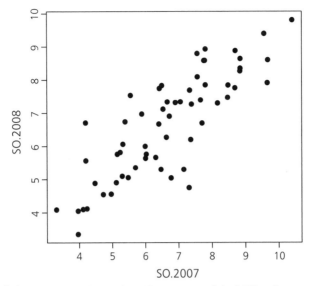

Figure 1.4. Scatterplot of the 2007 strikeout rate and the 2008 strikeout rate for all pitchers who started at least 25 games in both seasons. There is a relatively strong association pattern in the graph, indicating that the strikeout rate is a good measure of a pitcher's ability.

so a pitcher's strikeout rate,

$$SO/9 = 9 \times \frac{SO}{IP},$$

where SO and IP are respectively the counts of strikeouts and innings pitched, may be a better measure of a pitcher's ability. Figure 1.4 displays a scatterplot of the 2007 strikeout rate and the 2008 strikeout rate for our collection of starting pitchers. We see a strong relationship in the graph that indicates that strikeout rate is a better measure of a pitcher's ability than ERA; 69% of the variability of the 2008 strikeout rates can be explained by the 2007 strikeout rates.

A pitcher's strikeout rate is an example of a *defense independent pitching statistic* or $DIPS$, a measure of pitching performance that does not involve the fielders. Other examples of $DIPS$ are home runs allowed (HR), hit batters (HBP), and walks (BB). One pitching measure based only on these statistics is called the defense-independent component ERA or $DICE$:

$$DICE = 3.00 + \frac{13 \times HR + 3(BB + HBP) - 2 \times SO}{IP}.$$

Figure 1.5 shows a scatterplot of a pitcher's 2007 $DICE$ statistic with his

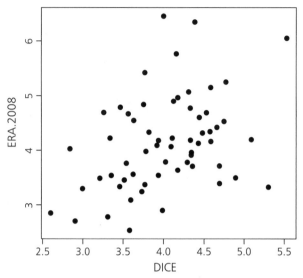

Figure 1.5. Scatterplot of the 2007 dependent independent pitching measure *DICE* and the 2008 *ERA* for all pitchers who started at least 25 games in both seasons. The 2007 *DICE* measure is slightly better than the 2007 *ERA* in predicting the following season's *ERA* for a pitcher.

2008 *ERA*. There is a stronger relationship between *DICE* and the 2008 *ERA* than the 2007 and 2008 *ERA*s (Figure 1.3). So, the defense-independent component measure *DICE* is a better predictor of the following year's *ERA* than the present year's *ERA*. It can be shown that 15 percent of the variability in the 2008 *ERA* can be explained by the 2007 *DICE* measure.

1.4 Measuring Fielding

Sabermetricians are also interested in finding good measures of fielding performance. The basic fielding measures are the counts of put-outs (PO), assists (A), and errors (E). The traditional measure of fielding ability is the fielding percentage,

$$FLD\% = \frac{PO + A}{PO + A + E}.$$

This statistic measures the proportion of fielding plays that were made successfully. However, this measure ignores balls outside of the reach of the fielder where a play was not made. It is important for a fielder to have good *range*, that is, the ability to move a long distance for the ball that is placed in-play, and the fielding percentage says nothing about a fielder's range.

A simple way of measuring a fielder's range is to consider the count of plays that are made. Fielders with more range tend to be involved in more plays. One range measure is the range factor per innings $RF/9$, which is the number of plays (assists and putouts) made for each nine innings:

$$RF/9 = 9 \times \frac{PO + A}{IP}.$$

Let us use these two fielding measures to compare two modern great short-stops Omar Visquel and Derek Jeter. Both have been awarded the Gold Glove award for fielding excellence as a shortstop; Visquel won the award in the nine seasons 1993 to 2001 and Jeter won the award for the three seasons 2004 through 2006. Figure 1.6 displays the fielding percentages of both players plotted as a function of their age. Smoothing curves help us see the general patterns in the graph. The fielding percentages increase from left to right, which means that both fielders became more successful in making plays as they matured. Visquel's fielding percentages are substantially higher than Jeter's, indicating that Visquel is more successful (by about one percent) in making plays. Figure 1.7 shows the range factors for both players plotted against age. A player's range tends to decrease with age, so older players are

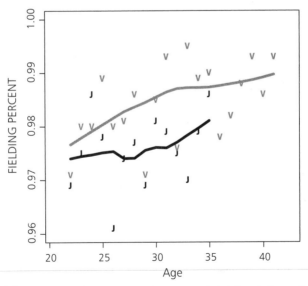

Figure 1.6. Fielding percentages of Omar Visquel and Derek Jeter plotted as a function of age where the plotting symbol is the first letter of the player's last name. Lowess smoothing curves are used to show the basic patterns in the percentages. It is clear that Visquel was more successful in making fielding plays.

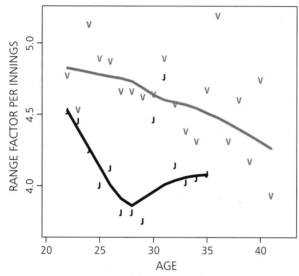

Figure 1.7. Range factors of Omar Visquel and Derek Jeter plotted as a function of age where the plotting symbol is the first letter of the player's last name. Lowess smoothing curves are used to show the basic patterns in the range factors. This graph shows that Visquel had a superior fielding range in his career.

less able to reach balls that are not hit directly to them. Also note that Visquel displays a much greater range than Jeter. Generally, Visquel is able to reach about a half additional play more than Jeter for each nine innings. This is a significant difference. Visquel is able to prevent more runs scored for his team than Jeter. In fact, Omar Visquel is generally considered to be one of the best fielding shortstops in baseball and Derek Jeter is considered (at least by sabermetricians) to be one of the worst fielding shortstops, despite the fact that he won three Gold Glove awards.

1.5 New Measurements, New Data and Measures of Performance

Baseball is continuously changing in its use of statistics. Major League Baseball (MLB) is measuring the game in new ways and the new data generated can help improve our measurements of player performance. During the 2007 baseball season, MLB began a systematic effort to record detailed information about the pitches that are thrown. All baseball stadiums were equipped with video cameras that would track each pitched ball and precisely determine its trajectory. From the measurements made from the cameras, one is able to learn about the speed of each pitch at its release point and at the point where it reaches home plate. Also one can measure the amount and angle of the "break" of the pitch. This technology is known as the PITCH F/X system.

Here is one illustration of the use of this new data to develop a new measure of pitching performance. One basic pitch in baseball is the fastball that is thrown at a high speed. A successful pitcher is able to throw the fastball at precise locations around the strike zone. A batter often will swing at a fastball. He will typically swing and miss at a "good" fastball thrown high in the strike zone, but be successful in connecting with a fastball that is thrown in the middle of the strike zone. By using the PITCH F/X data, we can measure the "sweet spot" where the batter has a high probability of connecting with the fastball. Figure 1.8 shows the location of the sweet spot in the strike zone for four pitchers, Paul Maholm, Cliff Lee, Johan Santana, and Edinson Volquez — these graphs are based on pitching data from the 2008 season. One sees that batters are generally successful in making contact with a fastball thrown low in the strike zone. The size of the sweet spot is an indication of the quality of the pitcher's fastball. Cliff Lee and Edinson Volquez threw

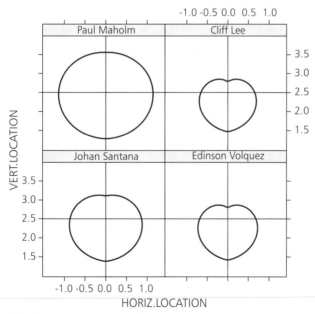

Figure 1.8. Contour plot of probability of making contact of a fastball for four pitchers. The line corresponds to the probability of 0.78; the region inside the contour line is the "sweet spot" for the batter where the probability of contact exceeds 0.78. The pitcher Edinson Volquez with an excellent fastball has a small sweet spot.

effective fastballs by keeping the batter's sweet spot small. In contrast, Paul Maholm had a less effective fastball and batters were able to make contact with balls pitched in a larger area.

MLB is currently testing a new measurement system that will record the exact speed and location of the ball and every player on the field during a game. This system will have many applications, but it will make it much easier to measure fielding ability. A fielder is responsible for covering a particular section of the playing field and this new system can record the success and failure of each play by the fielder in his zone. Allan Swartz in a *New York Times* article, says that this new system will allow "the most digitized of sports to be overrun anew by hundreds of innovative statistics that will rate players more accurately, almost certainly affect their compensation and perhaps alter how the game itself is played."

1.6 Further Reading

The book *Curve Ball* by Albert and Bennett provides a good introduction to sabermetrics and describes statistical thinking in the context of baseball. *Baseball Between the Numbers: Why Everything You Know About the Game Is Wrong* by Baseball Prospectus and *The Book: Playing the Percentages in Baseball* by Tango, Lichtman, and Dolphin describe a variety of sabermetrics topics such as clutch ability, baseball strategy, how much should players be paid, the impact of ballparks on statistics, and what statistics tell us about steroid use. The book *The Numbers Game: Baseball's Lifelong Fascination with Statistics* by Swartz gives a history of the use of statistics in baseball including some important people such as Henry Chadwick and Bill James.

About the Author

Jim Albert is a Professor of Mathematics and Statistics at Bowling Green State University. He is a Fellow of the American Statistical Association and past editor of *The American Statistician*. His research interests include Bayesian inference for categorical data, the teaching of statistics, and the application of statistical thinking in sports. He has written books on statistical thinking in baseball, Bayesian inference, and on the use of the R statistical system to perform Bayesian computations. Away from academia, Jim is an active tennis player and enjoys music and following the Philadelphia Phillies.

CHAPTER 2

Surprising Streaks and Playoff Parity: Probability Problems in a Sports Context

Rick Cleary

Abstract

Sporting events generate many interesting probability questions. In probability, coun-terintuitive results are common and even experienced mathematicians can make mis-takes in problems that sound simple on the surface. In this article we examine two sports applications in which probability questions have results that may come as a surprise to sports fans and mathematicians alike. The first concerns the chances of rare events in sporting contests, with recent examples from baseball and football. The second considers the merits of longer playoff series in sports such as major league baseball and professional basketball, where a set of games is used to determine who advances in a post-season tournament.

2.1 Problem 1: Rare Events

One of the most appealing aspects of watching sports events is the possi-bility that a viewer may see something especially noteworthy, perhaps even unprecedented. Sports fans who have the good fortune to be in attendance when something spectacular happens have vivid memories and wonderful stories. Baseball fans love to describe the time they saw a no-hitter, a player hit for the cycle, a triple play, or some other rare event. A typical response to such descriptions is, "Wow, what are the chances of that?"

Those of us who love sports and math frequently are asked, "What are the chances of . . . ?" It may be easy to come up with a quick estimate of the probability, but first guesses are often wrong, sometimes off by orders of magnitude! Making estimates of the probability of rare events can be difficult. Checking the answer may be hard because the events are sometimes so unusual that we do not have enough historical instances to get a sense of the probability. A problem of this type for sports fans occurs in the first round of the annual Division I men's basketball tournament of the National Collegiate Athletic Association. Since 1985 when the tournament field expanded to 64 (now 65) teams, a 16th seed has yet to beat a top-ranked first seed. The probability of a 16th seed winning a first round game is not zero, but it is small and we have no past history to inform us.

We look at examples of streaks in a baseball game and a football season as a way to explain some of the difficulties that occur when thinking about rare events.

Example 1: Four homers in a row

On April 22, 2007 the Boston Red Sox had a remarkable third inning against the New York Yankees. Red Sox slugger Manny Ramirez came to the plate with two out and nobody on and launched a home run. Then J. D. Drew, Mike Lowell, and Jason Varitek followed Ramirez and each of them hit a home run. Such a back to back to back to back event is rare indeed as this was only the fifth time in major league history that it had happened. And so on April 23 the always lively sports media conversations around Boston centered on it and the natural question was, "What are the chances of that?"

Many people were glad to try to answer. A Boston Globe sports reporter got estimates from several people with good credentials for this sort of work: a Red Sox employee, an academic from California who had looked into the question when the Los Angeles Dodgers hit four home runs in a row in a previous year, and a mathematics professor at one of the many colleges in the Boston area. (Author's note: It wasn't me!) According to the article one of the experts said:

> "What is needed in finding the various probabilities that you are interested in is the following ratio: $p = $ the total number of home runs hit in the major leagues, divided by the number of plate appearances.... It would be reasonable ... to use the numbers from 2006 to come up with that number. Last season, there were 5,386 home runs hit in 188,052 plate appearances. Thus p is equal to .02857. The probability of four consecutive home runs is p to the fourth power, or p times p times p times p. That equals .000000673, which in this case means there is a one in 1.4 million chance."

There is a problem with that estimate. In the approximately 170,000 major league baseball games that have taken place since 1900 the event "four home runs in a row" has happened five times. So history suggests a frequency like one in 34,000 rather than one in 1.4 million! What's going on?

Let's call the "1 in 1.4 million chance" our first estimate. Any baseball fan with a little bit of probability knowledge sees one problem and one questionable assumption in this estimate. The questionable assumption is that raising p to the fourth power assumes that the four events are independent, which may not be so. Baseball old-timers, for instance, would claim that the probability of a batter being hit by a pitch might go up significantly after two or three homers in a row, which would make the event less likely. Others would argue that seeing two or three homers in a row suggests that the pitcher or pitchers are struggling, which would make the event more likely. We retain the independence assumption for now because correcting it would require an empirical study of past games that is not our primary concern.

The problem is that the four hitters in our sequence are all power hitters whose home run probabilities are considerably higher than p. If we replace p to the fourth power by the product of the four individual home run rates the players had established in their careers through 2006, the estimate changes. The probabilities were .0608, .0403, .0369, and .0324 for Ramirez, Drew, Lowell, and Varitek respectively; all were veteran players so they are based on large samples. We can use the data to get a less naïve estimate of one in $(1/[(.0608)(.0403)(.0369)(.0324)]) = 340{,}814$. This indicates that four homers in a row by these players is about four times as likely as our first estimate. That's a big difference, but we're still off by a factor of ten from the historical data!

We now consider the difficulty from a probability modeling point of view. For a mathematician, asking "What are the chances of that?" requires that we define what "that" is. With our estimates we are computing the answer as if, when Manny Ramirez strolled to the plate, one fan turned to another and said, "What's the probability that we will see four home runs in a row right now?" Suppose instead that the same fans were having a conversation on the way into the ballpark and one asked the other, "What are the chances that we see four home runs in a row at any point during tonight's game?"

A typical major league baseball game has about 80 hitters come to the plate, so a streak of four homers has about 80 opportunities to get under way. For a rare event like this, the probability of its occurrence in 80 trials is very close to 80 times the probability that it occurs on any trial. (This logic fails for more common events because it doesn't take into account that the event may occur more than once.) Suppose we return to our first estimate for four typical major leaguers: One in 1.4 million becomes one in 1,400,000/80 or 1 in 17,500. This is now reasonably in line with the historical data.

There are lessons here for anyone hoping to use probability models. One is that we should always carefully define the event and the time frame. Another key lesson is to check, if possible, the estimate against the historically available data. Example 1 is a case in which respected academics and professionals, quoted in a leading newspaper, suggested a probability estimate that was not really incorrect, but it did not fit for the natural question. A problem about baseball is a good teaching tool and fun to consider, but we should think about how important it is to do a good job considering the probabilities of rare events outside of the sports pages. Estimates of the chances of financial market collapses, nuclear power plant failures, and exceptionally strong storms, just to name a few, are key drivers of public policy.

Example 2: A streak of winless opponents

At the start of the 2009 National Football League season, the Washington Redskins had a remarkable run. Their first six games were against winless opponents! What are the chances of that?

Here is a summary of a discussion of this problem that circulated among participants in a football pool (run for entertainment purposes only, of course!) made up of academics with PhDs in a variety of subjects. One contributor suggested that we might expect the probability of playing against six consecutive winless opponents to start the season would be about 1 in 32,768. This was determined by first computing the probability of playing six straight winless opponents, which was estimated to be $1 * (.50) * (.25) * (.125) * (.0625) * (.03125)$. The leading 1 appears because every team's opponent is winless in their first game. The reciprocal of the probability gives the 32,768.

This problem has many of the same features as in Example 1. A rare event is observed, and a probability is computed with an assumption of independence. As in the first example, where we replaced a home run ratio with player specific values, we can argue that the probabilities are not quite right. In fact, the Redskins week five opponent, the Carolina Panthers, had a bye week with no game in week four, which would change their probability of being winless in week five from .0625 to .125. The problem has another consideration that we should keep in mind: the event would be just as newsworthy if a team opened a season with six straight games against undefeated teams. Again, we find it hard to define the chances of an event until we have clearly defined the event.

Make the question more specific: What is the chance in any particular season that at least one National Football League team will begin playing six straight games against teams that are either all winless or all undefeated? Adding the undefeated teams, and taking into account that about one quarter of the teams will play a team that has had a bye in the first six weeks, makes some difference, but the most important change to the estimate of one in 32,768 comes from the fact that there are 32 teams in the league and if any of them had such a streak it would be in the news! (The 32 teams that might realize such a streak here correspond to the 80 at-bats where a home run streak might have started in Example 1.) My improved estimate is that the chance of the event is approximately once in every 400 seasons. Try your own calculation to see if you agree! The 2009 Redskins were in fact the first team to start the season with such a streak.

2.2   Problem 2: Playoff Series Length

Since 1995, major league baseball playoffs have consisted of eight teams. In each of the American and National leagues there are three divisions: East, Central, and West. The regular season champions in the six divisions advance to the playoffs along with one wild card team (the second place team with the best record) from each league. The first round of the playoffs, called the division series, consists of four best three-of-five games series between the playoff teams. The remaining two rounds, the two League Championship Series and the World Series, are best four-of-seven games series.

When an upset occurs in the division round, it is not hard to find fans of the losing team complaining that the shorter first round series was inherently and almost criminally unfair! Sports journalists agree. This is what Darren Everson and Hannah Karp wrote in the *Wall Street Journal* on October 6, 2009:

> Nothing else in sports is as thrillingly zany—or as patently ridiculous —as the first round of Major League Baseball's playoffs.
>
> After playing nearly 14 dozen games over six months (not including spring training), the best eight teams are paired up in a fiendish creation called the Division Series, where the first team to win three games moves on. There's really no logical way to defend this practice. It's as if the Boston Marathon chose the winner by putting its top finishers through a limbo contest.
>
> The Division Series, which begins Wednesday, has been so volatile in recent years that the team with the better record has won just 48%

of the time. When it comes to teams that win 100 games or more—
the mark of an elite season—only 10 of 19 have survived. In one year,
2002, no fewer than three 100-win teams went down in the first round.

"It's unfair," says Los Angeles Dodgers manager Joe Torre. "It's really Russian roulette."

There are few complaints about the fairness of seven game series. Are best
of seven game series really much more fair? Is it more likely that the superior
team will advance in a best of seven game series?

Suppose that teams A and B are about to meet in a playoff series and A is
the superior team. Let p represent the probability that A would defeat B in
a single game, so that $0.5 \leq p \leq 1$. Assume the outcomes of the games are
independent. Let w_5 and w_7 be the probabilities that A wins a best of five
and a best of seven series. Then

$$w_5 = p^3 + 3p^3(1-p) + 6p^3(1-p)^2.$$

The terms represent the probability of a three game series sweep by A, a
win by A in four games and a win by A in five games respectively. The
coefficient of 3 in the second term appears because there are three ways in
which B might have won one game (either by winning the first, second or
third game), while the 6 in the third term appears because there are six ways
to arrange two wins for A and two wins for B during the first four games.
For the longer series:

$$W_7 = p^4 + 4p^4(1-p) + 10p^4(1-p)^2 + 20p^4(1-p)^3.$$

If a seven game series is fairer than a five game series, we would expect that
w_5 and w_7 would differ widely for some values of p. But they do not, as
Figure 2.1 shows.

The maximum difference in these two curves occurs when p is about
0.689. At that point we find that $w_7 = 0.874$ and $w_5 = 0.837$, a difference of only 0.037. A value of $p = 0.689$ is larger than we would expect in
playoff games practice, and in fact it is extremely rare for a team to win 69%
of its games even during the regular season.

A criticism of this model is that p is not constant because, for one thing,
one of the teams will have home field advantage in each game. Also the
assumption that outcomes are independent is debatable. For example, a team
one loss away from elimination might use every possible strategic weapon
available to win a particular game, such as using a starting pitcher on short
rest, or having their best relief pitcher work longer than usual. Though we
can add these to our model, the curves in Figure 2.1 move very little.

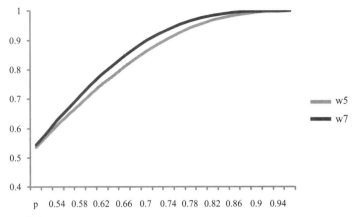

Figure 2.1. Probability that a team with probability p of winning an individual game wins a five game series (lower curve) or a seven game series (upper curve).

In the two problems we have studied, people who think carefully about sports and probability suggested estimates that were wrong. The rare events were unlikely, but not fantastically unusual. Longer playoff series favor the better team, but the differences are probably less than most fans believe. Sports are a wonderful source for probability problems, and probability can shed light on sports and challenge the conventional wisdom in interesting and entertaining ways. Students learning about probability would be wise to adopt the lessons in these examples.

References

[1] *Boston Globe*, April 24, 2007,
www.boston.com/sports/baseball/redsox/articles/2007/04/24/talk_about_long_shots/

[2] *Wall Street Journal*, October 6, 2009   online.wsj.com/article/SB10001424052748703298004574455520978204270.html

About the Author

Rick Cleary is Professor and Chair of the Department of Mathematical Sciences at Bentley University. He previously taught at Saint Michael's College and Cornell University. He received his PhD in statistics from Cornell. He specializes in applied statistical analyses. In the past few years he has worked on problems in sports, biomechanics, market research and plant pathology, among others. He likes participating in sports at least as much as he likes analyzing their mathematical components. In particular he enjoys running, golf, and basketball.

CHAPTER 3

Did Humidifying the Baseball Decrease the Number of Homers at Coors Field?

Howard Penn

Abstract

Because Coors Field in Denver is one mile above sea level, a baseball hit there will travel about 10% farther than one hit with the same force and angle at sea level. To partially nullify this effect in 2002 the Colorado Rockies began humidifying the balls used there. In this paper we show that doing so statistically reduced the number of home runs hit there.

Figure 3.1. ©Craig Welling www.rockiesphotos.blogspot.com/

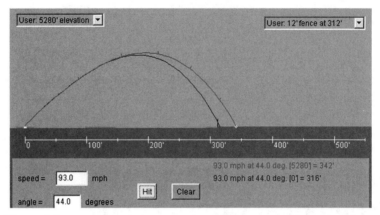

Figure 3.2. [5] ©James Carr faculty.tcc.fl.edu/scma/carrj/Java/baseball4.html

3.1 Introduction

Coors Field, where the Colorado Rockies play, has always been regarded as a home run friendly ballpark. It is at an altitude of approximately 5280 feet. According to the *Physics of Baseball* [1], a batted ball travels approximately 10% farther at that altitude than at sea level. Figure 3.2 compares the flight of two batted balls with the same initial speed, angle of elevation, and wind speed. The lower curve represents the path if the ball is hit at sea level and the upper curve shows the path at an altitude of 5280 feet. In the latter case the ball clears the fence but not in the former.

The Rockies were aware of this when Coors field was built. The ballpark has among the largest average dimensions of any major league ballpark. Figure 3.3 is a photo of opening day [4]

Table 3.1 shows a comparison of the distance and height of the fences at Coors with Safeco Field in Seattle, AT&T Ballpark in San Francisco, and Yankee Stadium (both the old and new Yankee Stadiums have identical dimensions) [3]. Safeco and AT&T are considered to be ballparks that favor pitchers.

Table 3.1.

Stadium	LF	HT	LC	HT	CF	HT	RC	HT	RF	HT
Coors Field	347	8	390	8	415	8	375	17	350	17
Safeco Field	331	8	390	8	405	8	386	8	326	8
AT&T (2004)	339	8	364	11	399	8	421	8	309	25
Yankee Stadium	318	8	399	7	408	13.83	353	14.5	314	10

Figure 3.3. ©Ballparks of Baseball www.ballparksofbaseball.com

In an attempt to compensate for the extra distance that batted balls fly, in 2002 the Rockies began humidifying the baseballs. They keep them in a room where the temperature is 70 degrees and the humidity is 50%, conditions that are similar to those at ballparks at sea level. The idea is that the humidity will make the baseballs slightly larger and softer, so they would not fly as far. Figure 3.4 shows Jay Alves, the vice president of marketing in the Humidor.

Figure 3.4. ©AP photo / Ed Andrieski

3.2 The Numbers

From 1999 to 2001, 816 home runs were hit at Coors Field (an average 272 per year). From 2002 to 2008, 1380 home runs were hit there [6, 7] (an average 197.1 per year), a decrease of nearly 75 home runs per year. Most papers about this [2] conclude that humidifying decreases the number of home runs. We claim that the analysis is not sufficient. The Rockies could have traded for better pitchers, giving up home run hitters in the exchange. Anyone who follows baseball would consider that unlikely. Maybe testing for performance enhancing drugs cut down the number of home runs throughout baseball. Perhaps the balls were different in the latter years. A deeper analysis is needed to determine whether the humidifying was the cause of the decrease in home runs.

3.3 A Useful Statistic

The possible factors can be accounted for by comparing the number of home runs hit in Rockies home games by both teams with the number hit in away games. Teams play 81 home and 81 away games. The opponents are, more or less, the same for both. (There are a few inter-league games where the Rockies play a team only at home or only away.) The opponent's pitchers and batters are the same and the Rockies have the same players for both home and away games. Therefore, if the ballpark and baseballs are homer neutral, then home runs should be just as likely to be hit at home or away. Hence the proportion of homers hit in home games should have a binomial distribution with mean $p = 0.5$. From 1999 to 2001 the number of home runs in Rockies away games was 478. This gives the percentage of home runs in home games as 63.06%.

We can now use the large sample test for population proportion [8]:

$$H_0 : p = p_0$$

$$z = \frac{\hat{p} - p_0}{\sqrt{\frac{p_0(1-p_0)}{n}}}$$

$$H_a : p > p_0 \text{ is } P(Z \geq z).$$

We are testing the hypothesis that $p_0 = 0.5$. The alternate hypothesis, H_a, is that the proportion of home runs at Coors Field is significantly more that 0.5. The computation of z is the normal approximation to the binomial distribution, whose value indicates how many standard deviations the data is above the assumed mean. The value of P is the probability that a set of random data will have a z value greater than or equal to the one computed. This represents the significance level of the hypothesis test. In statistics, the null hypothesis is usually rejected if P is less than 0.05. The data for 1999–2001 gives $z = 9.40$ and $P(Z \geq 9.40) < 10^{-10}$. This result is significant and we can conclude that before 2002 Coors Field was definitely a ball park that favored homers. The number of home runs for 2002–2008 in away games was 1178. So 53.95% were hit in home games. This data gives $z = 4.00$ and $P(Z \geq 4.00) \approx 0.000032$, which is still significant.

The data indicate that we can be virtually certain that Coors Field was a home run friendly ballpark before the team began humidifying baseballs. Although the percentage of home runs in home games has dropped remarkably, the ballpark remains a statistically significant home run friendly ballpark.

3.4 Comparing the two sets of data

The computations by themselves do not tell us if the drop in percentage is statistically significant. In order to determine that we may use the significance test for comparing two proportions [8]:

$$H_0 : p_1 = p_2$$
$$z = \frac{\hat{p}_1 - \hat{p}_2}{SE_{D_p}}$$

where the pooled standard error is

$$SE_{D_p} = \sqrt{\hat{p}(1 - \hat{p}) \left(\frac{1}{n_1} + \frac{1}{n_2} \right)}$$

where

$$\hat{p} = \frac{X_1 + X_2}{n_1 + n_2}$$
$$H_a : p_1 > p_2 \text{ is } P(Z \geq z).$$

With our data we get

$$z = \frac{\frac{816}{1294} - \frac{1386}{2558}}{0.0168882942} = 5.40$$

and

$$P(Z \geq 5.40) \approx 3.43 \times 10^{-8}.$$

The null hypothesis is that there is no significant difference between the percentages of home runs hit in home games. The alternate hypothesis is that the difference is significant and unlikely to have occurred by chance. The pooled standard deviation combines both sets of data, with $\hat{p}$ as the percentage of home runs in Rockies games at home over the entire time period. This data produces a significance level of 3.43×10^{-8}, which is so low that we can conclude that humidifying the baseball has decreased the number of home runs.

3.5 Summary of Conclusions

Before the Rockies began humidifying baseballs, Coors Field was an extraordinarily home run friendly ballpark. Since using the humidor, the ballpark remains home run friendly despite the drop in percentage. The drop is statistically significant.

The assumption that the away games are fair is not completely correct. The Rockies are in the same division as the San Francisco Giants and the San Diego Padres. Both of their home ballparks are statistically significantly difficult ballparks to hit home runs. Given the unbalanced schedule, this affects the percentage of home runs hit in the Rockies home games. At the same time, the fact that the other two teams play a good number of away games at Coors Field negatively affects *the percentage of the season total home runs that are hit in their home games.*

3.6 Exercises

Two teams with traditionally pitcher friendly ballparks have taken steps to increase the number of home runs in their teams home games. In 2004, the San Francisco Giants moved the center field fence at AT&T Park in from 404 feet to 399 feet. In 2003, the Detroit Tigers moved the left center field fence in from 395 feet to 370 feet [3]. Did that significantly increase the number of home runs? In 2009 the Yankees opened the new Yankee Stadium

Table 3.2. The data [9]

Team	Home	Away
Giants 1999–2003	574	861
Giants 2004–2008	665	779
Tigers 1999–2002	388	572
Tigers 2003–2008	1050	1166
Yankees 1999–2008	1898	1993

across the street from the old one. Its dimensions are exactly the same and the ballpark was designed to make it look like the old one. Yet the feeling among sportscasters is the number of home runs has increased. Is the new stadium statistically more homer friendly than the old one?

References

[1] R.K. Adair, *Physics of Baseball, 3rd Ed.* Harper-Collins, New York, NY, 2002.

[2] Associated Press, "Humidor Means Coors Field is no longer a hitter's heaven," available at www.bostonherald.com/sports/baseball/red_sox/view.bg?articleid=1040626

[3] Ballparks by Mausey and Suppes, "Dimensions of Ballparks," available at www.ballparks.com

[4] Ballparks of Baseball, "Photo of Coors Field," available at www.ballparksofbaseball.com

[5] J. Carr, "Tallahassee Community College, Baseball Simulator Java Applet," available at faculty.tcc.fl.edu/scma/carrj/Java/baseball4.html

[6] ESPN "Major League Baseball Statistics," available at www.sports.espn.com/mlb/statistics

[7] MLB.com, "MLB Statistics," available at www.mlb.com/mlb/stats/index.jsp

[8] D.S. Moore and G.P. McCabe, *Introduction to the Practice of Statistics, 3rd Ed.*, W.H. Freeman and Company, New York, NY, 2000

[9] H.L. Penn, "Table of home runs, 1999–2008," unpublished.

[10] Welling, Craig, "Photo of World Series Home Run," available at www.rockiesphotos.blogspot.com/

About the Author

Howard Penn received his B.A. in Mathematics from Indiana University in 1968 and his Masters and PhD in Mathematics from the University of Michigan in 1969 and 1973. Since receiving his final degree, he has taught at the United States Naval Academy where he held the rank of Professor. He was one of the pioneers is the use of computer graphics in the teaching of mathematics, having written the program, MPP, which was widely used internationally. He was also interested in finding applications of the mathematics to areas that many students will find interesting. Outside of mathematics, he was a well-known amateur photographer with special interests in wildlife, landscapes and abstract photography. His work may be viewed at www.flickr.com/photos/howardpennphoto

CHAPTER 4

Streaking: Finding the Probability for a Batting Streak

Stanley Rothman and Quoc Le

4.1 Introduction

In baseball, a player can gain instant fame by duplicating or exceeding one of the fabled types of batting streaks. The most well known is Joe DiMaggio's (1941) streak of hitting in 56 consecutive games. There are also other batting streaks such as Ted Williams' (1949) 84-game consecutive on-base streak, Joe Sewell's (1929) 115-game streak of not striking out in a game, and the 8-game streak of hitting at least one home run in each game, held by three players. Other streaks include most consecutive plate appearances with a hit (the record is 12 held by Walt Dropo (1952)), most consecutive plate appearances getting on-base (the record is 16 held by Ted Williams (1957)), and the most consecutive plate appearances with a walk (the record is seven held by five players).

In this paper, we present two functions to calculate the probability of a player duplicating a hitting streak. One is recursive; the other is a new piecewise function that calculates the probability directly.

We use them to compare the difficulty of duplicating different streaks, in particular, the 56-game consecutive hitting streak and the 84-game consecutive on-base streak.

Our results can be applied to streaks outside of baseball.

4.2 A recursive function to calculate the probability of a player having a 56-game hitting streak at some point in a season

Michael Freiman [1] gives a recursive function, $R(n)$, to calculate the probability of a player having a 56-game hitting streak at some point in the season.

Let
H = the number of hits in the season
PA = the number of plate appearances in the season.
A plate appearance will include an at-bat (AB), a walk (BB), a hit-by-pitch (HBP), a sacrifice fly (SF), and a sacrifice bunt (SH).
G = the number of games a player had at least one plate appearance in the season,
$p = H/PA$,
$j.kl = PA/G$ (for $j.kl = 5.18$, j is 5 and $.kl$ is $.18$),
$U = [1 - (1 - p)^j] * (1 - .kl) + [1 - (1 - p)^{(j+1)}] * .kl$,
$R(0) = R(1) = R(2) = \cdots = R(55) = 0$,
$R(56) = U^{56}$,
$R(n) = R(n - 1) + (1 - R(n - 57)) * (1 - U) * U^{56}$; when $n > 56$.

The average number of plate appearances per game is $j.kl$, the probability of getting at least one hit in a game for a given season is U, and the probability of getting a hit in any plate appearance is p. U^{56} is the probability of having a 56-game hitting streak in any particular 56-game span. The value $R(n)$ is the probability of having at least one 56-game hitting streak in the first n games.

The function $R(n)$, assumes the following:

1. The probability p is assigned to each plate appearance in the season.

2. The outcomes of plate appearances are independent events.

3. The player has the same number of plate appearances in each game for the season, $j.kl$.

4. A player's batting performance in each game is independent of his performance in any other game.

Some baseball researchers have questioned the independence assumptions and have attempted to show the number of streaks predicted from them underestimate the actual number of streaks. For this paper, these assumptions will be used.

$R(n)$ is the probability of a player having at least one 56-game hitting streak at some point in the first n-games in a season.

At this point, we explain the logic used to define $R(n)$.

For a player to have a 56-game hitting streak, at some point, in the first n-games, he must either have a 56-game hitting streak in the first $(n-1)$ games or have his first hitting streak in the last 56 games. The probability of having a 56-game hitting streak at some point in the first $(n-1)$ games is $R(n-1)$. To have his first hitting streak in the last 56 games, three events must all occur. First, he must not have a 56-game hitting streak in the first $(n-57)$ games, then he must not have a hit in game number $(n-56)$, finally, he must get a hit in each of the games $(n-55)$ through n. Call the three events, E_1, E_2, and E_3:

$E_1 = $ Not having a 56-game hitting streak in the first $(n-57)$ games,

$E_2 = $ Not having a hit in game number $(n-56)$,

$E_3 = $ Getting a hit in each of the games from game number $(n-55)$

through game number n.

$$Pr(E_1 \text{ and } E_2 \text{ and } E_3) = Pr(E_1) * Pr(E_2) * Pr(E_3)$$
$$= (1 - R(n-57)) * (1 - U) * U^{56}$$

The event of having a 56-game hitting streak at some point in the first $(n-1)$ games and the compound event of $(E_1 \text{ and } E_2 \text{ and } E_3)$ are mutually exclusive (they both cannot occur at the same time). Therefore,

$$R(n) = R(n-1) + (1 - R(n-57)) * (1 - U) * U^{56}$$

Next, a non-recursive function, that we developed to evaluate $R(n)$ is given.

4.3 A non-recursive piecewise function, $NR(n)$, to calculate the probability of a player having a 56-game hitting streak at some point in a season

$NR(n)$ is the probability of at least one hitting streak of length 56 in the first n games. We will call $NR(n)$ the NR **Function** to distinguish it from the **Recursive Function** $R(n)$.

The NR Function for calculating the probability of a player achieving a 56-game hitting streak at some point in the first n-games of a season is defined as follows.

For $0 \leq n \leq 55$, $NR(n) = 0$.

For $56 \leq n \leq 112$, $NR(n) = U^{56} + (n - 56) * (1 - U) * U^{56}$.

For $113 \leq n \leq 162$, $NR(n) = U^{56} + (n - 56) * (1 - U) * U^{56} -$
$.5 * (n - 112) * (1 - U) * U^{112} * [(n - 112) * (1 - U) + 1 + U]$.

The meaning of U is the same as in the recursive function.

To see the advantage of the NR Function over the recursive function, consider the following example.

Example 1   Suppose that during a season player A played in 66 games with $AB = 322, H = 138, BB = 20, HBP = 0, SF = 0$, and $SH = 0$. What is the probability that at some point in his season he had a 56 game hitting streak?

Solution 1   Using the recursive function

Step 1: $p = H/PA = 138/(322 + 20 + 0 + 0 + 0) = .404$ (p is the probability of a hit in a plate appearance).

Step 2: The average number of plate appearances per game expressed as a decimal is $j.kl = PA/G = 342/66 = 5.18$. ($j = 5, .kl = .18$)

Step 3: The weighted average, U, for the probability of a player getting at least one hit in $j.kl$ plate appearances per game is

$$U = \left[1 - (1 - p)^{j}\right] * (1 - .kl) + \left[1 - (1 - p)^{(j+1)}\right] * .kl$$
$$= \left[1 - (1 - .404)^{5}\right] * (1 - .18) + \left[1 - (1 - .404)^{(5+1)}\right] * .18$$
$$= \left[1 - (.596)^{5}\right] * (.82) + \left[1 - (.596)^{6}\right] * (.18)$$
$$= .9248 * .82 + .9552 * .18$$
$$= .9302$$

Step 4: The probability of a player having a 56-game hitting streak in any particular 56-game span in his season of 66 games is

$$U^{56} = .9302^{56} = .0174.$$

Step 5: The use of a recursive function begins with a known beginning term.

The known beginning term is $R(56) = U^{56}$. Our goal is to reach $R(66)$.

To reach it we must find $R(57), R(58), \ldots, R(66)$.

$$R(0) = R(1) = R(2) = \; = R(55) = 0,$$
$$R(56) = U^{56} = .9302^{56} = .0174,$$
$$\begin{aligned} R(57) &= R(56) + (1 - R(n - 57)) * (1 - U) * U^{56} \\ &= R(56) + (1 - R(0)) * (1 - .9302) * (.9302)^{56} \\ &= .0174 + .0012 = .0186, \end{aligned}$$
$$\begin{aligned} R(58) &= R(57) + (1 - R(n - 57)) * (1 - U) * U^{56} \\ &= R(57) + (1 - R(1)) * (1 - U) * U^{56} = .0186 + .0012 = .0198, \end{aligned}$$
$$\begin{aligned} R(59) &= R(58) + (1 - R(n - 57)) * (1 - U) * U^{56} \\ &= R(58) + (1 - R(2)) * (1 - U) * U^{56}.0198 + .0012 = .0210, \end{aligned}$$
$$\begin{aligned} R(60) &= R(59) + (1 - R(n - 57)) * (1 - U) * U^{56} \\ &= .0210 + .0012 = .0222. \end{aligned}$$

Continuing, we obtain

$$R(61) = .0234, \; R(62) = .0246, \; R(63) = .0258,$$
$$R(64) = .0270, \; R(65) = .0282, \; \text{and} \; R(66) = .0292.$$

Solution 2   Using the NR Function

Since $56 \leq n \leq 112$,

$$NR(n) = U^{56} + (n - 56) * (1 - U) * U^{56}.$$
$$NR(66) = (.9302)^{56} + (66 - 56) * (1 - .9302) * (.9302)^{56}.$$
$$NR(66) = .0292.$$

The functions $R(n)$ and $NR(n)$ can be adjusted for any s-game hitting streak as follows:

For the **recursive function** $R(n)$, we define

$$R(0) = R(1) = R(2) = \cdots = R(s - 1) = 0, \quad R(s) = U^s$$

and

$$R(n) = R(n - 1) + \left(1 - R(n - (s + 1))\right) * (1 - U) * U^s; \quad \text{where } n > s.$$

For the **piecewise function** $NR(n)$, we

let n be the number of games played by a player in a season.

$NR(n)$ is the probability of at least one s-game hitting streak at some point in the first n games.

For $0 \leq n \leq (s - 1)$, $NR(n) = 0$.

For $s \leq n \leq 2s$, $NR(n) = U^s + (n - s) * (1 - U) * U^s$.

For $(2s + 1) \leq n \leq (3s + 1)$,

$$NR(n) = U^s + (n - s) * (1 - U) * U^s - .5 * (n - 2s) * (1 - U)$$
$$* U^{2s} * \left[(n - 2s) * (1 - U) + 1 + U \right]$$

The generalized NR Function, $NR(n)$, is valid only when the number of the first n-games is less than or equal to $3s + 1$.

If s is replaced in the above function by 56, we obtain the function for the probability of having a 56-game hitting streak. **Since the number of games in a current season is equal to 162 and prior to 1961 was 154 games, the NR Function can be applied to both Joe DiMaggio's 56-game hitting streak and to Ted Williams' 84-game on-base streak or to any streak such that $(3s + 1) \geq 162$. The NR Function can also be used if the number of the first n-games is less than or equal to $3s + 1$.** For streaks with smaller lengths s, if the number of games n exceeds $3s + 1$, $NR(n)$ must be defined differently for each of the intervals $[3s + 2, 4s + 2]$, $[4s + 3, 5s + 3]$, $[5s + 4, 6s + 4]$, ...

For $n > (3s + 1)$, the definition of $NR(n)$ for n is dependent on which of the above intervals contain n. Instead of defining a new function for each of them, it turns out that if we use the function $NR(n)$ defined for n in the interval $[2s + 1, 3s + 1]$ in most cases the error will be very small.

We will now examine the error in using $NR(n)$, defined for n in the interval $[2s + 1, 3s + 1]$, when, in fact, $n > (3s + 1)$. The letter n represents the first n-games a player played in a given season. We define the error by the absolute value of the difference between $R(n)$ and $NR(n)$ $(Error = |R(n) - NR(n)|)$.

4.4   The Error $= |R(n) - NR(n)|$

The Error Table provides, for a given streak length from $s = 5$ to $s = 52$, the maximum value for U that would result in an error less than $1 * 10^{-4}$. For a given streak length, s, any U value less than or equal to the U value in Table 4.1 would also result in an error less than $1 * 10^{-4}$. Before 1961 a season consisted of 154 games. Starting in 1961, a season consisted of 162 games. Table 4.1 shows how changing from a season with 154 games to one with 162 games affects the value of U. The interpretation of this is if a player plays in fewer games a larger U can be tolerated. The U that works for $n = 162$

Table 4.1. The Error Table

Streak Length	$n = 162$	$n = 154$	Streak Length	$n = 162$	$n = 154$
S	U	U	S	U	U
5	0.2374	0.2403	29	0.8412	0.8449
6	0.3076	0.3108	30	0.8484	0.8522
7	0.3701	0.3734	31	0.8552	0.8590
8	0.4249	0.4284	32	0.8616	0.8655
9	0.4729	0.4764	33	0.8677	0.8717
10	0.5150	0.5186	34	0.8735	0.8776
11	0.5521	0.5557	35	0.8791	0.8833
12	0.5848	0.5884	36	0.8844	0.8888
13	0.6139	0.6175	37	0.8896	0.8941
14	0.6399	0.6435	38	0.8945	0.8992
15	0.6632	0.6667	39	0.8993	0.9043
16	0.6842	0.6877	40	0.9040	0.9092
17	0.7032	0.7067	41	0.9086	0.9140
18	0.7205	0.7240	42	0.9130	0.9189
19	0.7363	0.7398	43	0.9175	0.9237
20	0.7507	0.7542	44	0.9219	0.9287
21	0.7640	0.7675	45	0.9263	0.9338
22	0.7762	0.7797	46	0.9307	0.9392
23	0.7876	0.7911	47	0.9353	0.9452
24	0.7981	0.8016	48	0.9401	0.9521
25	0.8079	0.8114	49	0.9453	0.9611
26	0.8170	0.8206	50	0.9511	0.9772
27	0.8256	0.8292	51	0.9581	
28	0.8337	0.8373	52	0.9679	

games will also work for n less than 162 games. Also, as s increases, a larger U can be tolerated.

Therefore, if a player misses a few games, in a season, consisting of 162 games, the table value for U would still be valid.

If we choose a particular player who plays more than $3s + 1$ games and his value for U is less than or equal to the U in the table, we know that the error in using $NR(n)$ will be less than $1 * 10^{-4}$.

In the next example, we look at the record streak of hitting a home run in

Table 4.2.

Year	Player	#G	#PA	#HR	U	$NR(n)$ Function	$R(n)$ Function
1956	Dale Long	148	582	27	0.1702	0.0000826	0.0000826
1987	Don Mattingly	141	629	30	0.1956	0.00002313	0.00002313
1993	Ken Griffey Jr	156	691	45	0.2575	0.0021428	0.0021428

eight consecutive games. This has been accomplished by three players. Dale Long did it in 1956, Don Mattingly in 1987, and Ken Griffey Jr. in 1993. For this streak, Step 1 in Example 1 becomes $p = HR/PA$ and Step 4 in Example 1 becomes U^8. For Dale Long (1956) we have:

$$p = 27/582 = .046; \quad j.kl = 582/148 = 3.93; \quad s = 8; \quad U = .1702.$$

$$NR(n) = U^s + (n - s) * (1 - U) * U^s - .5 * (n - 2s) * (1 - U)$$
$$* U^{2s} * \left[(n - 2s) * (1 - U) + 1 + U \right].$$
$$NR(148) = (.1702)^8 + (148 - 8) * (1 - .1702) * (.1702)^8 - .5$$
$$* (148 - 16) * (1 - .1702) * (.1702)^{16}$$
$$* \left[(148 - 16) * (1 - .1702) + 1 + .1702 \right]$$
$$= .0000826.$$

Dale Long had a 1 in 12,048 chance of achieving his streak in 1956.

Using both the $NR(n)$ and $R(n)$ functions, Table 4.2 provides the probability of each player achieving this streak in the year that they did. Table 4.1 shows for $s = 8$ and $U < 0.425$ we would expect these two probabilities to differ by no more than $1 * 10^{-4}$. Clearly, the three players who hold the record all have a small enough value for U to satisfy the error condition. From Table 4.2, the two probabilities agree to seven decimal places. Table 4.2 shows that, because Griffey was a better home run hitter than Long, he was 26 times more likely to hit a home run in eight consecutive games.

4.5   Generalizing the concept of a streak

In this section, we expand the concept of a streak to other streaks in and outside of baseball.

4.5.1   Definitions

A **binomial experiment** consists of k **trials**, $t_1, t_2, t_3, \ldots, t_k$. Each trial results in one of two mutually exclusive outcomes (one outcome is called success; the other outcome is called failure). The trials are independent and the probability of success is the same for each trial.

Success is associated with one of the two outcomes of a trial; **failure** is associated with the other outcome.

A **season** consists of a fixed number, N, of trials. The first trial is called trial number 1 and the last trial is trial number N.

Let W equal the number of successes in the season.

A **game** consists of a certain number of trials in a season. Game #1 begins with trial number 1 and ends at a trial less than or equal to N. Game #2 begins with the first trial after the last trial of game #1. The games continue in this manner with the last game ending after the last trial. Every trial belongs to exactly one game. The number of trials in each game can be the same or can be different. A game can consist of exactly one trial.

The **number of games in a season** is $m = G$.

The length of the season is N trials and consists of m games. The games are sequential with the first trial appearing in game #1 and the last trial appearing in game number m.

Season $= t_1, t_2, t_3, \ldots, t_N$.   ($t_i$ is either a success or failure)

Game 1 $= t_1, t_2, \ldots, t_k$

Game 2 $= t_{(k+1)}, t_{(k+2)}, \ldots, t_{(k+z)}$

Game 3 would start at $t_{(k+z+1)}$.

$\downarrow$

Game m contains trial t_N as the last trial.

The **probability of success** for each trial is $p = W/N$. The **number of trials used for each game** is $j.kl = N/m$. (j is the integer part; $.kl$ is the decimal part.) **The probability of at least one success in a game is**

$$U = \left(1 - (1 - p)^j\right) * (1 - .kl) + \left(1 - (1 - p)^{(j+1)}\right) * .kl.$$

A streak of length s means having at least s consecutive games with at least one success in each game.

We assume independence between individual trials as well as between games. Even though individual games need not have the same number of trials, for our model, we will use, $j.kl$, the average number of trials per game for a season, for each game. Also, the same p will be used for each trial in each game in the season.

The following inputs and calculations are needed to evaluate the probability of a player duplicating a given streak.

4.5.2   Inputs and calculations

Inputs:

$$N = \text{number of trials in a season}$$
$$W = \text{number of successes in a season}$$
$$m = \text{number of games in a season}$$
$$s = \text{length of the streak}$$

Calculations:

$$p = W/N \quad \text{[probability of a success in a typical trial]}$$

$$N/m = j.kl \quad \text{[number of trials in a typical game]}$$

$$U = (1 - (1 - p)^j) * (1 - .kl) + (1 - (1 - p)^{(j+1)}) * .kl$$
$$\text{[probability of at least one success in a typical game]}$$

$NR(n)$ is equal to the probability of at least one streak of length s in the first n-games. The piecewise function, $NR(n)$, was defined on page 35.

When each plate appearance is considered a game (that is, $N = m$), we can look at streaks involving individual plate appearances. Since $N = m$ and $j.kl = N/m$, we have $j.kl = 1.00$, $j = 1$, $.kl = .00$, and $U = p$. The next two examples look at this case.

4.5.3   Each Individual plate appearance is a game

Example 2   The current record of 12 consecutive plate appearances with a hit is held by Walt Dropo (1952). Pinky Higgins (1938) had 12 consecutive hits in 12 consecutive at-bats. Higgins actually needed 14 plate appearances to achieve his 12 consecutive hits. One can say that if we looked at 12 consecutive hits in 12 consecutive at-bats these two players would share the record. Example 4 examines this streak.

Both these players are unlikely candidates to hold the record. Higgins, a career .292 hitter, batted .303 in 1938. Dropo, a career .270 hitter, batted .276 in 1952. We will use the NR function, $NR(n)$, to find the probability for Walt Dropo achieving this streak. A plate appearance is a trial; each plate appearance is a game; success is getting a hit in a plate appearance. Since $m > 3s + 1$ for these two players we provide in Table 4.3 an error table for $s = 12$ and $m = 162$, $m = 500$, $m = 600$, and $m = 700$. Since the error $|NR(n) - R(n)|$ decreases as s increases, Table 4.3 can be used for any $s \geq 12$.

Walt Dropo (1952 season):

$$N = PA = 633; \quad W = H = 163; \quad m = G = 633; \quad s = 12$$
$$p = W/N = 163/633; \quad j.kl = N/m = 633/633 = 1.00;$$
$$U = \left(1 - (1-p)^j\right) * (1 - .kl) + \left(1 - (1-p)^{(j+1)}\right) * .kl$$
$$= p = .257504$$
$$NR(n) = U^s + (n-s) * (1-U) * U^s - .5 * (n-2s) * (1-U)$$
$$* U^{2s} * \left[(n-2s) * (1-U) + 1 + U\right]$$

$NR(633)$ and $R(633)$ give the probability of Dropo having 12 hits in 12 consecutive plate appearances.

$$NR(633) = .000039276$$
$$R(633) = .000039275$$
$$Error = |R(633) - NR(633)| = .000000001.$$

Since $m < 700$ and $U < 0.50$, from Table 4.3, we would expect the error, $|R(633) - NR(633)|$ to be less than 0.0001.

The $NR(n)$ function shows Walt Dropo's chance of achieving his streak of 12 consecutive hits in 12 consecutive plate appearances in 1952 is 1 in 25,641.

Example 3 The current record (since 1900) for consecutive plate appearances getting on base is 16 held by Ted Williams in 1957. For the 1957 season we have:

$$N = PA = 546;$$
$$W = OB = H + BB + HBP = 163 + 119 + 5 = 287;$$
$$m = G = 546; \quad s = 16$$
$$p = W/N = .525641; \quad j.kl = N/m = 546/546 = 1.00$$
$$U = p = .525641$$
$$NR(546) = .0085385$$
$$R(546) = .0085386$$
$$Error = |R(546) - NR(546)| = .0000001$$

The $NR(n)$ function shows that Ted Williams' chance of getting on base in 16 consecutive plate appearances is 1 in 117.

Table 4.3. Error Table for $s = 12$

U	$m = 162$	$m = 500$	$m = 600$	$m = 700$
0.50			0.0001	0.0001
0.51		0.0001	0.0001	0.0002
0.52		0.0001	0.0002	0.0003
0.53		0.0002	0.0004	0.0006
0.54		0.0004	0.0007	0.0011
0.55		0.0007	0.0012	0.0019
0.56		0.0012	0.0021	0.0034
0.57		0.0021	0.0037	0.0059
0.58	0.0001	0.0036	0.0063	0.0101
0.59	0.0001	0.0061	0.0106	0.0171
0.60	0.0002	0.0102	0.0178	0.0285
0.61	0.0004	0.0168	0.0294	0.0468
0.62	0.0006	0.0274	0.0477	0.0757
0.63	0.0010	0.0440	0.0764	0.1206
0.64	0.0016	0.0698	0.1205	0.1892
0.65	0.0026	0.1092	0.1872	0.2925
0.66	0.0041	0.1682	0.2868	0.4453
0.67	0.0065	0.2554	0.4326	0.6676
0.68	0.0100	0.3824	0.6430	0.9858
0.69	0.0153	0.5642	0.9414	1.4336
0.70	0.0230	0.8203	1.3581	2.0537

4.5.4 Each Individual At-Bat is a Game

Example 4   As mentioned in Example 2, the record for consecutive hits in consecutive at-bats is 12 hits in 12 at-bats. This record is shared by Walt Dropo (1952) and Pinky Higgins (1938). For this streak, a plate appearance which is not an at-bat is considered a game in which the player did not play. For this reason N = AB is used instead of N = PA.

Walt Dropo (1952 season):

$$N = AB = 591; \quad W = H = 163; \quad m = G = 591; \quad s = 12$$
$$p = W/N = 163/591; \quad j.kl = N/m = 591/591 = 1.00;$$
$$U = (1 - (1 - p)^j) * (1 - .kl) + (1 - (1 - p)^{(j+1)}) * .kl$$
$$= p = .275804$$
$$NR(n) = U^s + (n - s) * (1 - U) * U^s - .5 * (n - 2s) * (1 - U)$$
$$* U^{2s} * [(n - 2s) * (1 - U) + 1 + U]$$

$NR(591)$ and $R(591)$ give the probability of Dropo having 12 hits in 12 consecutive at-bats.

$$NR(591) = .0000814240$$
$$R(591) = .0000814243$$
$$Error = |R(591) - NR(591)| = .0000000003.$$

This translates into a 1 in 12,281 chance of Dropo achieving this streak in 1952.

Pinky Higgins (1938 season)

$$N = AB = 524; \quad W = H = 159; \quad m = G = 524; \quad s = 12$$
$$p = W/N = 159/524; \quad j.kl = N/m = 524/524 = 1.00;$$
$$U = (1 - (1 - p)^j) * (1 - .kl) + (1 - (1 - p)^{(j+1)}) * .kl$$
$$= p = .303435$$
$$NR(n) = U^s + (n - s) * (1 - U) * U^s - .5 * (n - 2s) * (1 - U)$$
$$* U^{2s} * [(n - 2s) * (1 - U) + 1 + U]$$

$NR(524)$ and $R(524)$ give the probability of Higgins having 12 hits in 12 consecutive at-bats.

$$NR(524) = .00021787$$
$$R(524) = .00021786$$
$$Error = |R(524) - NR(524)| = .00000001.$$

This translates into a 1 in 4,590 chance of Higgins achieving this streak in 1938.

4.6 Comparing Ted Williams' 84-game consecutive on-base streak to Joe DiMaggio's 56-game consecutive hitting streak

In 1941, two great players performed fabulous batting feats. Joe DiMaggio had his 56-game hitting streak and Ted Williams was the last major league player to bat over .400 for a season. During his 56-game hitting streak, Joe DiMaggio had 91 hits in 223 at-bats for an average of .408. In 1941, Ted Williams had a 23-game hitting streak, the only year that Ted had a hitting streak of at least 20 games. In 1949, Williams established the record of getting on base in 84 consecutive games. In 1941, wrapped around Joe DiMaggio's 56-game hitting streak, was a streak of 74 consecutive games getting on base. Getting on base means either reaching base by a hit, or a walk, or being hit by a pitch. Reaching base on an error or a fielder's choice does not count. Table 4.4 shows the batting statistics for DiMaggio (1941) and Williams (for the 1941 season and 1949 season). A success in a given plate appearance is different for the two streaks. For the 56-game streak, success is getting a hit; for the 84-game streak, success is getting on base. For the 56-game streak, $p = (H/PA)$. For the 84-game streak, $p = (OB/PA)$; where OB represents the total number of plate appearances in the season which resulted in the player getting on-base.

One question that is often asked is: Which of these two streaks would be hardest to duplicate? If the two streaks involved the same number of games the answer would be obvious. Since a player's season on base total is greater than or equal to his season hit total, the probability of a success in a plate appearance would be higher with the on-base streak. This would lead to a higher value for U and a higher $NR(n)$ for the on-base streak. However, since the streak of 84 consecutive games is 28 more games than DiMaggio's 56-game hitting streak, this might lead to the belief that the 84-game streak would be harder to duplicate.

From Table 4.4, DiMaggio had a probability of 0.0001 of achieving his streak, whereas, Williams in 1949 had a probability of 0.09444 of achieving his streak. **For every 10,000 seasons, we would have expected DiMaggio in 1941 to accomplish his streak once. For every 10,000 seasons, we would have expected Williams in 1949 to accomplish his streak 944 times.** In 1941, DiMaggio's ratio of the probability of the 84-game streak to the probability of the 56-game streak was 56.70. This indicates that, for the year DiMaggio achieved his 56-game hitting streak, he was actually 56 times likelier to have the 84-game consecutive on-base streak. In fact, in 1941, Williams' probability of achieving his 84-game consecutive on-base streak was 0.15700.

Comparing Williams' two streaks of 16 consecutive plate appearances getting on base (Example 3) and his 84-game consecutive game on-base streak, his probability of .09444 of achieving his 84-game streak in 1949 was 11 times likelier than his probability of .00854 of achieving his 16 consecutive plate appearance getting on-base streak in 1957. **Williams had a 1 in 11 chance of achieving his 84-game streak in 1949 and a 1 in 117 chance of achieving his 16 consecutive on-base streak.**

Table 4.4. The 1941 Season

Player	Year	G	PA	AB	H	BB	BA	OBP	HBP	SH	SF	$NR(n)$ Prob. 56 Streak	$NR(n)$ Prob. 84 Streak	Ratio 84 Streak 56 Streak
DiMaggio	1941	139	621	541	193	76	0.357	0.440	4	0		0.00010	0.00565	56.50
Williams	1941	143	606	456	185	147	0.406	0.553	3	0		0.00002	0.15700	7850.00
Williams	1949	155	730	566	194	162	0.343	0.490	2	0		0.00001	0.09444	13491.43

Table 4.4 supports DiMaggio's streak as the harder one to duplicate. It is apparent that the real difference between the two players is their number of walks. A walk in a game extends the consecutive on-base streak but has a negative effect on the consecutive hitting streak. If a player walks it uses up a plate appearance.

We will now look at other great hitters to compare their probabilities of achieving these two streaks.

4.7 These two streaks evaluated for other great hitters

In the history of baseball, one of the greatest 5-year spans of hitting belongs to Rogers Hornsby. For the years from 1921 to 1925 his cumulative batting average was .402. Table 4.5 shows Hornsby's statistics for these five years. Hornsby had a 33-game hitting streak in 1922. From Table 4.5, we can see that in 1922 Hornsby had his highest probability of achieving the 56-game hitting streak.

Using the ratio of the probability of Hornsby having an 84-game on-base streak to the probability of him having a 56-game hitting streak, we see the ratio went from a low of 4 to a high of 47. For each year, Hornsby had a higher probability of achieving the 84-game consecutive on-base streak. Hornsby's statistics provide further evidence that DiMaggio's 56-game streak would be harder to duplicate.

Tables 4.6 and 4.7 present the batting statistics for two players recognized by many baseball people as the two best hitters since 2000. These players are Ichiro Suzuki and Albert Pujols.

Table 4.5. Rogers Hornsby

Player	Year	G	PA	AB	H	BB	BA	OBP	HBP	SH	SF	$NR(n)$ Prob. 56 Streak	$NR(n)$ Prob. 84 Streak	Ratio 84 Streak 56 Streak
Hornsby	1921	154	674	592	235	60	0.397	0.458	7	15		0.00121	0.00763	6.31
Hornsby	1922	154	704	623	250	65	0.401	0.459	1	15		0.00356	0.01475	4.14
Hornsby	1923	107	487	424	163	55	0.384	0.459	3	5		0.00053	0.00762	14.38
Hornsby	1924	143	640	536	227	89	0.424	0.507	2	13		0.00225	0.05804	25.80
Hornsby	1925	138	605	504	203	83	0.403	0.489	2	16		0.00043	0.02052	47.72

Table 4.6. Ichiro Suzuki

Player	Year	G	PA	AB	H	BB	BA	OBP	HBP	SH	SF	$NR(n)$ Prob. 56 Streak	$NR(n)$ Prob. 84 Streak	Ratio 84 Streak 56 Streak
Suzuki	2001	157	738	692	242	30	0.350	0.381	8	4	4	0.00116	0.00055	0.47
Suzuki	2002	157	728	647	208	68	0.321	0.388	5	3	5	0.00003	0.00065	21.67
Suzuki	2003	159	725	679	212	36	0.312	0.352	6	3	1	0.00004	0.00003	0.75
Suzuki	2004	161	762	704	262	49	0.372	0.414	4	2	3	0.00360	0.00523	1.45
Suzuki	2005	162	739	679	206	48	0.303	0.350	4	2	6	0.00001	0.00003	3.00
Suzuki	2006	161	752	695	224	49	0.322	0.370	5	1	2	0.00011	0.00024	2.18
Suzuki	2007	161	736	678	238	49	0.351	0.396	3	4	2	0.00052	0.00085	1.63
Suzuki	2008	162	749	686	213	51	0.310	0.361	5	3	4	0.00003	0.00008	2.67

Ichiro Suzuki throughout his baseball career in Japan was recognized as a superstar. He led the Japanese League in batting average for each year he played in Japan. In 2001, at the age of 27, he began his career in the United States. In each of his first eight years, he had over 200 hits. Ichiro is a lead-off hitter who had over 725 plate appearances for each year. He received relatively few walks. He batted from the left side. Even though Suzuki hit home runs, he was known more as a singles hitter with great speed who could beat out many infield singles. Before 2009, Ichiro had logged five hitting streaks of at least 20 games, which tied him with Ty Cobb who also had five 20+ hitting streaks. The record is seven 20+ hitting streaks held by Pete Rose. In 2001, Ichiro had a 23-game hitting streak and a 21-game hitting streak. In 2004, he had a 21-game streak; in 2006, he had a 20-game streak; and in 2007, he had a 25-game streak. From Table 4.6, we see that for the years 2001, 2004, 2006, and 2007, Ichiro had his highest probabilities of achieving a 56-game hitting streak. For the year 2004, Ichiro had his highest probability of achieving both streaks. What we find interesting is that for the years 2001 and 2003 his probability of achieving the 56-game hitting streak was higher than his probability of achieving the 84-game on-base streak. Looking at the ratio of the probability of the 84-game streak to the probability of the 56-game streak, for all but one year it was at or below three.

Albert Pujols started his career in 2001 at the age of 21. He was nicknamed

Table 4.7. Albert Pujols

Player	Year	G	PA	AB	H	BB	BA	OBP	HBP	SH	SF	$NR(n)$ Prob. 56 Streak	$NR(n)$ Prob. 84 Streak	Ratio 84 Streak 56 Streak
Pujols	2001	161	676	590	194	69	0.329	0.403	9	1	7	0.000004	0.00028	70.00
Pujols	2002	157	675	590	185	72	0.314	0.394	9	0	4	0.000002	0.00023	115.00
Pujols	2003	157	685	591	212	79	0.359	0.439	10	0	5	0.000060	0.00482	80.33
Pujols	2004	154	692	592	196	84	0.331	0.415	7	0	9	0.000010	0.00198	198.00
Pujols	2005	161	700	591	195	97	0.330	0.430	9	0	3	0.000004	0.00295	738.50
Pujols	2006	143	634	535	177	92	0.331	0.431	4	0	3	0.000005	0.00328	656.00
Pujols	2007	158	679	565	185	99	0.327	0.429	7	0	8	0.000002	0.00218	1090.00
Pujols	2008	148	641	524	187	104	0.357	0.462	5	0	8	0.000011	0.01129	1026.45

"the machine" due to his great hitting ability. In contrast to Suzuki, Pujols batted from the right side and is known as a power hitter. Albert was feared by pitchers, which led to his receiving many walks each season. He batted either in the third spot or fourth spot in the batting order. Except for two years, his number of plate appearances for a season was in the upper 600s to 700. The last column of Table 4.7 shows that the ratio of his probability of achieving an 84-game on-base streak to his probability of achieving a 56-game hitting streak ranged from 70 to over 1000.

Both players provide us with more evidence favoring the 56-game hitting streak as the streak that is the more difficult to duplicate.

Tables 4.8 and 4.9 will examine all players, since 1900, with the longest consecutive hitting streak and the longest consecutive on-base streak.

In Table 4.8, every ratio of the probability of an 84-game on-base streak to the probability of a 56-game hitting streak is greater than 1.80. This shows that even for those players with the longest consecutive hitting streaks, there was a higher probability of achieving the 84-game consecutive on-base streak.

In Table 4.9, every ratio of the probability of an 84-game on-base streak to the probability of a 56-game hitting streak is greater than 6.33. This shows that for the players with the longest consecutive on-base streaks, their probability of achieving the 84-game consecutive on-base streak was considerably higher than their probability of achieving the 56-game consecutive hitting

Table 4.8. Longest Consecutive Game Hitting Streaks

Streak Length	Player	Year	G	PA	AB	H	BB	BA	OBP	HBP	SH	SF	$NR(n)$ Prob. 56 Streak	$NR(n)$ Prob. 84 Streak	Ratio 84 Streak 56 Streak
56	DiMaggio	1941	139	620	541	193	76	0.357	0.440	4	0		0.000105	0.005650	53.81
44	Rose	1978	159	729	655	198	62	0.302	0.362	3	2	7	0.000010	0.000074	7.40
41	Sisler	1922	142	654	586	246	49	0.420	0.467	3	16		0.010570	0.019085	1.81
40	Cobb	1911	146	654	591	248	44	0.420	0.467	8	11		0.008430	0.015280	1.81
39	Molitor	1987	118	542	465	164	69	0.353	0.44	2	5	1	0.000072	0.004460	61.94
37	Holmes	1945	154	713	636	224	70	0.352	0.420	4	3		0.000326	0.004140	12.70
36	Rollins	2005	158	732	677	196	47	0.290	0.338	4	2	2	0.000005	0.000012	2.40

Table 4.9. Longest Consecutive Game On-Base Streaks

Streak Length	Player	Year	G	PA	AB	H	BB	BA	OBP	HBP	SH	SF	$NR(n)$ Prob. 56 Streak	$NR(n)$ Prob. 84 Streak	Ratio 84 Streak 56 Streak
84	Williams	1949	155	730	566	194	162	0.343	0.490	2	0		0.000007	0.094446	13492.29
74	DiMaggio	1941	139	620	541	193	76	0.357	0.440	3	0		0.000105	0.005650	53.81
69	Williams	1941	143	606	456	185	147	0.406	0.553	3	0		0.000024	0.157000	6541.67
65	Williams	1948	137	638	509	188	126	0.369	0.490	3	0		0.000067	0.081327	1213.84
63	Cabrera	2006	153	675	607	171	51	0.282	0.335	3	3	11	0.000000	0.000002	6.33
58	Snider	1954	149	679	584	199	84	0.341	0.423	4	1	6	0.000041	0.003813	92.55
58	Bonds	2003	130	550	390	133	148	0.341	0.529	10	0	2	0.000000	0.073201	3660050
57	Boggs	1985	161	758	653	240	96	0.368	0.450	4	3	2	0.000587	0.024690	42.06

streak.

Comparing Table 4.8 to Table 4.9 we observe

- Joe DiMaggio is the only player in both tables.

- Ted Williams appears three times in Table 4.9.

- Excluding Joe DiMaggio, every player in Table 4.8 had fewer than 71 walks.

- Excluding Joe DiMaggio and Orlando Cabrera, every player in Table 4.9 had at least 84 walks. Orlando Cabrera's appearance in Table 4.9 is unusual considering he had only 51 walks in 2006. Cabrera had a 1 in 500,000 chance of achieving the 84-game on-base streak.

Tables 4.8 and 4.9 provide further evidence that the 56-game hitting streak is more difficult to duplicate than the 84-game consecutive on-base streak.

We now turn to some streaks that involve more than one success in each game. We give two examples next.

Consecutive game streaks *with more than one success* in each game:

Most consecutive games with two or more hits in each game — The record is 13 held by Rogers Hornsby (1923).

Most consecutive games with three or more hits in each game — The record is 6 held by two players (since 1900). The two players were George Brett (1976) and Jimmy Johnston (1923).

Define:

$$p = H/PA$$

$$j.kl = PA/G$$

[$j.kl$ is the average number of plate appearances (trials) used for each game]

$x =$ the minimum number of successes in each game of the streak

$$U = [Pr(X \geq x), \text{ for } j \text{ trials}] * (1 - .kl)$$
$$+ [Pr(X \geq x), \text{ for } (j + 1) \text{ trials}] * .kl$$

[since the assumptions for this model satisfy the Binomial Model, the two probabilities can be calculated using the Binomial Formula]

$_mC_k =$ the number of combinations of m distinct objects taken k at a time.

We now look at two cases for a minimum number of x successes in a game, based on a fixed number of trials. A trial corresponds to a plate appearance.

Case 1 If $x = 1$:

$Pr(X \geq 1$, for j trials per game)

$\quad = 1 - Pr(X \leq (x - 1))$

$\quad = 1 - Pr(X \leq (1 - 1))$; for j trials)

$\quad = 1 - Pr(X = 0) = 1 - {}_j C_0 * p^0 (1 - p)^j$

$\quad = (1 - (1 - p)^j)$

$Pr(X \geq 1$, for $(j + 1)$ trials per game)

$\quad = 1 - Pr(X \leq (x - 1))$

$\quad = 1 - Pr(X \leq (1 - 1))$; for $(j + 1)$ trials)

$\quad = 1 - Pr(X = 0) = 1 - {}_{(j+1)} C_0 * p^0 (1 - p)^{(j+1)}$

$\quad = (1 - (1 - p)^{(j+1)}) \quad [{}_j C_0 = 1$ and ${}_{(j+1)} C_0 = 1]$

Up to now, the only streaks studied required $x = 1$ (at least one success in each game of the streak). This is what is used in the calculation for U in the 56-game hitting streak and all other streaks discussed before in this paper.

Case 2 If $x > 1$:

$Pr(X \geq x$, for j trials per game) $= 1 - Pr(X \leq (x - 1))$,

$\quad$ for j trials)

$Pr(X \geq x$, for $(j + 1)$ trials per game) $= 1 - Pr(X \leq (x - 1))$,

$\quad$ for $(j + 1)$ trials)

You can either use a Binomial Table or the Binomial Formula in Excel to calculate $Pr(X \leq (x - 1))$.

We will look at the example of Rogers Hornsby's 13-game consecutive streak of at least two hits in each game which he accomplished in 1923. From Table 4.5, the statistics for Hornsby in 1923 were

$$G = 107, \quad PA = 487, \quad H = 163$$
$$p = 163/487 = .335, \quad j.kl = 487/107 = 4.55$$
$$U = [Pr(X \geq 2), \text{ for } j = 4 \text{ trials}] * (1 - .55)$$
$$+ [Pr(X \geq 2), \text{ for } j = 5 \text{ trials}] * (.55)$$
$$U = [1 - Pr(X \leq 1), \text{ for } j = 4 \text{ trials}] * (1 - .55)$$
$$+ [1 - Pr(X \leq 1), \text{ for } j = 5 \text{ trials}] * (.55)$$

Using the binomial formula in Excel, we calculated the above probabilities. The resulting value for U was .482416. Using .482416 for U, the probability of Hornsby achieving the 13-game streak of at least two hits in each game for the 1923 season was $R(107) = .0038003$ and $NR(107) = .0038003$.

From Table 4.5, the probability of Hornsby achieving the 56-game hitting streak in 1923 was $NR(107) = .00053$. **Hornsby had a 1 in 263 chance of achieving the 13-game streak and a 1 in 1,886 of achieving the 56-game streak.**

The ratio of .0038003 to .00053 is over 7. The interpretation is, in 1923, Hornsby had seven times the chance of duplicating his 13-game streak than duplicating the 56-game hitting streak.

4.8 Conclusion

Two functions were provided to calculate the probability of various players duplicating various streaks in baseball. The two functions are a recursive function named $R(n)$ and a direct piecewise function called $NR(n)$. Both functions have for their domains the first n games of a player's season. The piecewise function $NR(n)$ is easier to use. However, the $NR(n)$ function is correct only when n is less than or equal to $3s + 1$ where s is the length of the streak. Since a player's season consists of 162 games or less, for such streaks as Joe DiMaggio's 56-game hitting streak and Ted Williams' 84-game consecutive on-base streak, the $NR(n)$ function will equal the $R(n)$ function.

For streaks with a small length s, even though the $NR(n)$ function has an error when $n > 3s + 1$, for many streaks this error is very small.

Two categories of streaks were discussed. The first category was streaks based on games; the second category was streaks based on plate appearances. Ted Williams had record streaks of both types. He holds the record for consecutive games getting on base at least one time (84 games in 1949). He also holds the modern day record of consecutive plate appearances getting on base (16 times in 1957). Using the $NR(n)$ and $R(n)$ functions, it was shown that Williams' consecutive plate appearance on-base streak would be harder to duplicate than the 84-game streak, based on the years in which Williams accomplished these two streaks.

Much of this chapter compares DiMaggio's 56-game hitting streak to Williams' 84-game consecutive on-base streak. The probability of achieving each of these two streaks is calculated for several great hitters of the past and present. The results indicated that of these two streaks DiMaggio's

56-game hitting streak would be the hardest to duplicate.

The paper also looks at game streaks involving more than one success in each game. The probability of achieving such streaks as Rogers Hornsby's consecutive 13-game streak of getting at least two hits in each game is calculated.

We end this paper with these questions. As mentioned in this paper, for streaks with small lengths and $n > 3s + 1$, each of the intervals $[3s + 2, 4s + 2], [4s + 3, 5s + 3], [5s + 4, 6s + 4], \ldots$ requires its own function $NR(n)$. Can you produce a piecewise function that will work for each of these intervals? Another question is in comparing the 56-game consecutive hitting streak to the 84-game consecutive on-base streak, can we find a function to calculate for any player, how many games must be added to 84 games or subtracted from 84 games so that the probability of that player achieving both streaks is the same? Finally, we ask for which independent variables is the rate of change greatest for the function

$$NR(n, U, s) = U^s + (n - s) * (1 - U) * U^s - .5 * (n - 2s) * (1 - U)$$
$$* U^{2s} * \left[(n - 2s) * (1 - U) + 1 + U \right];$$

$0 \leq U \leq 1, 2s + 1 \leq n \leq 3s + 1$, and $s \geq 5$?

As mentioned in the introduction, we are looking for applications, outside the world of baseball, which can apply the models for streaks developed in this paper.

References

[1] Michael Freiman, 56-Game Hitting Streaks Revisited, *The Baseball Journal*, SABR, 31 (2003) 11–15.

About the Authors

Stanley Rothman received his BS from Montclair State University in 1966 and his MS and Ph.D. from the University of Wisconsin in 1967 and 1970. His entire teaching and research career has been spent at Quinnipiac University. From 1970 to 1977, he was an Assistant Professor of Mathematics, from 1978 to 1984, he was an Associate Professor of Mathematics. Since 1985, he has been a full professor of mathematics. From 1992 he has served as Chairman of the Mathematics Department. In 1973, he founded a computer consulting firm called Acc-U-data. He served as President from 1973 to 2005. Since 2006, his major research interest has been in Sabermetrics. He spoke at both the San Diego and Washington DC National Mathematics Conferences in 2008 and 2009. The theme of both talks was the teaching of statistics from baseball records. He is now finishing his new book *Basic Statistics: Using Baseball to Bring Statistics to Life* which will be published by Johns Hopkins Press in 2010.

Quoc Le is a senior joint major in mathematics and finance at Quinnipiac University. He is in the process of applying to various graduate schools. His future plans include obtaining a Ph.D. and teaching at the university level.

Part II

Basketball

CHAPTER 5

Bracketology: How can math help?

Tim Chartier & Erich Kreutzer & Amy Langville & Kathryn Pedings

Abstract

Every year, people across the United States predict how the field of 65 teams will play in the Division I NCAA Men's Basketball Tournament by filling out a tournament bracket for the postseason play. This article discusses two popular rating methods that are also used by the Bowl Championship Series, the organization that determines which college football teams are invited to which bowl games. The two methods are the Colley Method and the Massey Method, each of which computes a ranking by solving a system of linear equations. The article also discusses how both methods can be adapted to take late season momentum into account. All the methods were used to produce brackets in 2009 and their results are given, including a mathematically-produced bracket that was better than 97% of the nearly 4.5 million brackets submitted to ESPN's Tournament Challenge.

5.1 Introduction

Every year around the beginning of March, students and faculty at 65 schools along with a large portion of the United States get excited about the prospects of March Madness, the Division I NCAA Men's Basketball Tournament. At the start of the tournament each of the 65 teams has a chance of ending the season crowned the champion. With the excitement of watching basketball comes the adventure of attempting to predict the outcome of the tournament by filling out a bracket. The NCAA estimates that 10% of the nation will fill out a bracket [4]. Be it participating in a small contest with friends or submitting your bracket to the nationwide ESPN Tournament Challenge, everyone wants to do well. How can we use mathematics to help? That is the goal of this paper—to describe several mathematical methods that can improve your chances of winning the office sports pool and possibly even the $1 million prize for the ESPN Tournament Challenge.

Sports teams are often ranked according to winning percentage, which is easily calculated for the ith team as $p_i = w_i/t_i$, where w_i is the total number of wins for team i and t_i is the total number of games it played. To make our computations concrete, we will examine data from the 2009 season of the Southern Conference of Division I NCAA Men's Basketball. The standings at the end of the regular season are in Table 5.1. Ranking the teams by winning percentage produces the results seen in the table.

While Davidson College is ranked first according to winning percentage, the team's two regular season conference losses were to the College of

Table 5.1. The regular 2009 season standings and winning percentage ratings for the teams in the Southern Conference of Division I NCAA Men's Basketball.

Team	Record	Rating	Rank
Davidson College	18–2	0.90	1
College of Charleston	15–5	0.75	2
The Citadel	15–5	0.75	2
Wofford College	12–8	0.60	4
Univ. of Tennessee at Chattanooga	11–9	0.55	5
Western Carolina Univ.	11–9	0.55	5
Samford Univ.	9–11	0.45	7
Appalachian State Univ.	9–11	0.45	7
Elon Univ.	7–13	0.35	9
Georgia Southern Univ.	5–15	0.25	10
Univ. of North Carolina at Greensboro	4–16	0.20	11
Furman Univ.	4–16	0.20	11

Charleston and the Citadel. Should Davidson still be ranked first or should the losses change the ranking? Davidson's winning percentage would be the same whether one of its losses came from second place College of Charleston or last place Furman University. Should the College of Charleston and the Citadel be tied for second place just because their winning percentages are the same? Could examining against whom teams won and lost serve to predict teams' future performance better?

These issues are reasons why many rating systems incorporate more than winning percentages. This article discusses two popular rating methods used by the Bowl Championship Series, the organization that determines which college football teams are invited to which bowl games [5]. The two methods are the Colley method and the Massey method. The Colley method was created by astrophysicist Wesley Colley who saw problems with ranking by winning percentage and developed a new method [1]. The Massey method started as Ken Massey's undergraduate honors math project and eventually made its way into the BCS [3]. The main idea behind this method is to use a least-squares approximation to find a rating vector.

Each method calculates ratings for all teams that can then be used to complete a bracket for the March Madness tournament by choosing the higher rated team as the winner for each matchup in the bracket. The goal of this paper is to examine if a mathematical strategy to completing a bracket can outperform the typical sports fan's bracket.

5.2 Colley Method

Ranking teams by winning percentage is a common ranking method even in professional sports. Wesley Colley proposed applying Laplace's rule of succession, which transforms the standard winning percentage into

$$r_i = \frac{1 + w_i}{2 + t_i}. \tag{5.1}$$

This minor change may appear to be of little help, but Colley used it as a stepping stone to a more powerful result. Before making such a step, let's apply this formula to the season's data in Table 5.1. Using (5.1) gives the ratings in Table 5.2. The rankings of the teams do not change but now the average of the ratings is 0.5, which we will now use in the derivation of the Colley method.

An alternative way to write the number of wins for a team is

$$w_i = \frac{w_i - l_i}{2} + \frac{w_i + l_i}{2} = \frac{w_i - l_i}{2} + \frac{t_i}{2},$$

Table 5.2. The regular 2009 season standings and ratings as calculated with (5.1) of the teams in the Southern Conference of Division I NCAA Men's Basketball.

	Team	Record	Rating	Rank
	Davidson	18–2	0.864	1
	Charleston	15–5	0.727	2
	Citadel	15–5	0.727	2
	Wofford	12–8	0.591	4
	Chattanooga	11–9	0.545	5
	W. Carolina	11–9	0.545	5
	Samford	9–11	0.455	7
	App. State	9–11	0.455	7
	Elon U	7–13	0.364	9
	Georgia Southern	5–15	0.273	10
	UNC Greensboro	4–16	0.227	11
	Furman	4–16	0.227	11

and

$$t_i/2 = \frac{1}{2}\overbrace{(1+1+\cdots+1)}^{\text{total number of games}} = \left(\frac{1}{2}+\frac{1}{2}+\cdots+\frac{1}{2}\right).$$

At the beginning of the season, all ratings are $1/2$, and they hover around $1/2$ as the season proceeds. So,

$$\frac{1}{2}\,(\text{total games}) \approx (\text{sum of opponents' ranks for all games played}),$$

or

$$\frac{1}{2}t_i \approx \sum_{j\in O_i} r_j,$$

where O_i is the set of opponents for team i. Substituting this back into our equation for w_i we get

$$w_i \approx \frac{w_i - l_i}{2} + \sum_{j \in O_i} r_j. \tag{5.2}$$

This substitution is approximate as the average over all opponents' ratings may not be 1/2 since, for one thing, every team may not play every other team.

Assuming equality in (5.2) and inserting this into (5.1) produces

$$r_i = \frac{1 + (w_i - l_i)/2 + \sum_{j \in O_i} r_j}{2 + t_i}. \tag{5.3}$$

The advantage of this representation is that we now have the interdependence of our teams' ratings. That is, team i's rating depends on the ratings r_j of all its opponents. This procedure for computing ratings is called the Colley method.

While each r_i can be computed individually using (5.3), an equivalent formulation uses a linear system

$$C\mathbf{r} = \mathbf{b}, \tag{5.4}$$

where C is the so-called Colley matrix. How is this linear system derived from (5.3)? It is easiest to see the contributions of (5.3) if it is written as

$$(2 + t_i)r_i - \sum_{j \in O_i} r_j = 1 + \frac{1}{2}(w_i - l_i). \tag{5.5}$$

The vector $\mathbf{b}$ has components $b_i = 1 + \frac{1}{2}(w_i - l_i)$. The diagonal elements of the Colley coefficient matrix C are $2 + t_i$ and the off-diagonal elements c_{ij}, for $i \neq j$, are $-n_{ij}$, where n_{ij} is the number of games between teams i and j.

Let's simplify by taking a subset of the Southern Conference and rank only Davidson, Charleston, the Citadel, Wofford, and Chattanooga according to their 2009 regular season records against each other. Their performance results in the linear system

$$\begin{pmatrix} 10 & -2 & -2 & -2 & -2 \\ -2 & 9 & -2 & -2 & -1 \\ -2 & -2 & 9 & -2 & -1 \\ -2 & -2 & -2 & 10 & -2 \\ -2 & -1 & -1 & -2 & 8 \end{pmatrix} \begin{pmatrix} r_1 \\ r_2 \\ r_3 \\ r_4 \\ r_5 \end{pmatrix} = \begin{pmatrix} 3 \\ 1.5 \\ 0.5 \\ 0 \\ 0 \end{pmatrix},$$

where r_1, r_2, r_3, r_4 and r_5 are the ratings for Davidson, Charleston, the Citadel, Wofford and Chattanooga, respectively. Solving this linear system yields Colley ratings of $r_1 = 0.667, r_2 = 0.554, r_3 = 0.464, r_4 = 0.417$ and $r_5 = 0.398$, which maintains the property that the average of the ratings is 1/2. The Colley matrix provides matchup information. For example, the (2,5)-entry of C means that team 2, Charleston, played team 5, Chattanooga, only once while Wofford played each of its opponents twice that season.

To emphasize interdependence, let's add a fictional game to this linear system. Suppose Chattanooga and Charleston play one more time with Chattanooga winning. Then the linear system becomes

$$\begin{pmatrix} 10 & -2 & -2 & -2 & -2 \\ -2 & 10 & -2 & -2 & -2 \\ -2 & -2 & 9 & -2 & -1 \\ -2 & -2 & -2 & 10 & -2 \\ -2 & -2 & -1 & -2 & 9 \end{pmatrix} \begin{pmatrix} r_1 \\ r_2 \\ r_3 \\ r_4 \\ r_5 \end{pmatrix} = \begin{pmatrix} 3 \\ 1 \\ 0.5 \\ 0 \\ 0.5 \end{pmatrix}.$$

The new ratings are $r_1 = 0.667, r_2 = 0.500, r_3 = 0.458, r_4 = 0.417, r_5 = 0.458$. The additional game affects the ratings of more than just the two teams playing, yet the property that the average ratings are 1/2 is maintained.

Let's return to the regular season results for the entire Southern Conference. Solving the 12×12 linear system produces the ratings in Table 5.3. The ranking from winning percentages are listed for comparison.

The Colley method is unaffected by differences in final scores. Nowhere is the game score considered. In the Colley method, a win is a win regardless of the score. The Massey method is a ranking method that includes game scores in the ratings.

5.3 Massey Method

If Davidson beat Elon by 10 points and Elon beat Furman by 5 points, could we then predict that Davidson would beat Furman by 15 points? If this were true, sports wouldn't be as much fun to watch but it would make sports ranking easy. Transitivity will rarely, if ever, hold perfectly, but assuming that it holds approximately is the foundation of the Massey method. Let r_1, r_2 and r_3 be the ratings for Davidson, Elon, and Furman and let's compute these ratings so that they can predict outcomes of future games. For our small example, $r_1 - r_2 = 10$. (Davidson beat Elon by 10 points.) The difference in the ratings r_1 and r_2 is the margin of victory achieved by the higher rated team. We also have $r_2 - r_3 = 5$. (Elon beat Furman by 5 points.) If we add

Table 5.3. The regular 2009 season standings and the Colley ratings of teams in the Southern Conference of Division I NCAA Men's Basketball.

Team		Record	Colley Rating	Colley Rank	Winning % Rank
🦁	Davidson	18–2	0.82	1	1
🐦	Charleston	15–5	0.74	2	2
♟	Citadel	15–5	0.68	3	2
🐺	Wofford	12–8	0.57	4	4
C	Chattanooga	11–9	0.56	5	5
🐱	W. Carolina	11–9	0.53	6	5
🐶	Samford	9–11	0.47	7	7
⚔	App. State	9–11	0.46	8	7
🔥	Elon U	7–13	0.40	9	9
🦅	Georgia Southern	5–15	0.28	10	10
🏀	UNC Greensboro	4–16	0.26	11	11
🐾	Furman	4–16	0.25	12	11

these two linear equations, we create the prediction $r_1 - r_3 = 15$. (Davidson beats Furman by 15 points.) How do we compute the ratings r_1, r_2 and r_3?

For this small example, we have the linear equations $r_1 - r_2 = 10$ and $r_2 - r_3 = 5$; there are infinitely many solutions: for any nonnegative integer c, $r_1 = 15 + c, r_2 = 5 + c$ and $r_3 = c$ is a solution. In practice, the situation is reversed. There are typically more games than teams, so we have a system with more equations than variables that in general will have no solution. For instance, suppose we add an additional game in which Furman beats Davidson by 1 point. Thus, $r_3 - r_1 = 1$. Along with our earlier linear equations this forms the linear system, $M_1 \mathbf{r} = \mathbf{p}_1$ or

$$\begin{pmatrix} 1 & -1 & 0 \\ 0 & 1 & -1 \\ -1 & 0 & 1 \end{pmatrix} \begin{pmatrix} r_1 \\ r_2 \\ r_3 \end{pmatrix} = \begin{pmatrix} 10 \\ 5 \\ 1 \end{pmatrix}.$$

You might think that a system of three equations and three unknowns would have a unique solution. However, regardless of the number of games played m between n teams, the matrix M_1 of dimensions $m \times n$ never has full rank. Because of the symmetry of matchups, any set of $n - 1$ columns can be used to build the remaining column, which means that the rank of M_1 is at most $n - 1$. Since we cannot find an exact solution to the system, we will find an approximate solution. We will use the method of least squares to find the vector $\mathbf{r}$ such that the length of the vector $\mathbf{p}_1 - M_1\mathbf{r}$, which is the residual error, is minimized.

To do this, we first compute $M_1^T M_1 \mathbf{r} = M_1^T \mathbf{p}_1$, which we denote as $M_2\mathbf{r} = \mathbf{p}_2$,

$$\begin{pmatrix} 2 & -1 & -1 \\ -1 & 2 & -1 \\ -1 & -1 & 2 \end{pmatrix} \begin{pmatrix} r_1 \\ r_2 \\ r_3 \end{pmatrix} = \begin{pmatrix} 9 \\ -5 \\ -4 \end{pmatrix},$$

where $M_2 = M_1^T M_1$ and $\mathbf{p}_2 = M_1^T \mathbf{p}_1$ is the cumulative point differential vector. Once again, $M_2\mathbf{r} = \mathbf{p}_2$ is a singular system since it can be proven that the rank of M_2 is $n - 1$. To create a nonsingular system $M\mathbf{r} = \mathbf{p}$, we need only take one additional step and replace a row in the matrix M_2 by a row of 1s to form M, the Massey matrix, and the corresponding entry in the vector $\mathbf{p}_2$ with a 0 to form $\mathbf{p}$. This step implies that the sum of the ratings should be 0. Solving our new linear system,

$$\begin{pmatrix} 2 & -1 & -1 \\ -1 & 2 & -1 \\ 1 & 1 & 1 \end{pmatrix} \begin{pmatrix} r_1 \\ r_2 \\ r_3 \end{pmatrix} = \begin{pmatrix} 9 \\ -5 \\ 0 \end{pmatrix},$$

produces the desired ratings. This method of rating is called the Massey method. While any row of M_2 can be replaced by a row of ones, replacing the last row is the convention established by Ken Massey in his undergraduate honors thesis for Bluefield College in Virginia. Solving this linear system we find that $r_1 = 3, r_2 = -1.667$, and $r_3 = -1.333$. Therefore, our method predicts Davidson (team 1) will beat Elon (team 2) by $3 - (-1.667) = 4.667$ points.

As with (5.4) for the Colley method, we can create the Massey linear system directly from a season's data. In all but the last row, the diagonal elements of the Massey matrix are the number of games played by team i and the off-diagonal elements m_{ij} are the negative of the number of games played between teams i and j. Notice that $C = M_2 + 2I$, which is convenient when building the two systems' ratings. The vector $\mathbf{p}_2$ is the sum of the point differentials in team i's games where a win gives a positive differential and a loss, a negative one.

Having seen the Massey method on a small example, let's return to the regular season results for the entire Southern Conference. We again have a 12×12 linear system to solve and this system, created according to the Massey method, results in the ratings found in Table 5.4. While both methods produce ratings that are interdependent, there are differences in the rankings. Chattanooga drops from fifth to eighth when ratings are produced by the Massey method. Given the interdependence of the ratings, such a drop can be difficult to fully explain. Part of the explanation is Chattanooga's big losses to highly ranked teams, including a 22 point loss to Davidson.

Table 5.4. The regular 2009 season standings and the Massey ratings of teams in the Southern Conference of Division I NCAA Men's Basketball.

	Team	Record	Massey Rating	Massey Rank	Colley Rank
	Davidson	18–2	13.99	1	1
	Charleston	15–5	5.42	2	2
	Citadel	15–5	5.32	3	3
	Wofford	12–8	0.73	5	4
	Chattanooga	11–9	-1.30	8	5
	W. Carolina	11–9	2.67	4	6
	Samford	9–11	0.00	6	7
	App. State	9–11	-0.17	7	8
	Elon U	7–13	-4.22	9	9
	Georgia Southern	5–15	-7.73	11	10
	UNC Greensboro	4–16	-6.57	10	11
	Furman	4–16	-8.14	12	12

5.4   Weighting Methods

Neither the Colley nor Massey method takes into account when in the season
a game occurs. This can be important. A star player may be injured half
way through the season or a team may mature and improve over time. It
seems especially valuable for March Madness predictions to boost the rating
of a team that has won its last ten games even if it performed poorly at
the beginning of the season. There are ways to adapt both methods to take
momentum into account.

In the standard version of the Massey and Colley methods, a game at the
beginning of the season contributes to a team's rating with the same weight
as a game at the end of the season. What if we want to reward teams for
playing well in the weeks leading up to the tournament? We can do this by
weighting games according to the date they were played, so the outcome of
recent games affects the ratings more than earlier ones.

5.4.1   Linear weighting and the Colley method

To weight the games, we want a function $g(t)$ that takes as input t, the number of days into the season at which the game occurs, and returns a real
positive weight as output. A simple weighting function is a linear weighting

$$g(t) = (t - t_0)/(t_f - t_0) = t/t_f, \tag{5.6}$$

where $t_0 = 0$ represents opening day for the season and t_f is the total number of days in the season. The weights in (5.6) for games played by Davidson
College against opponents in the Southern Conference are plotted in Figure
5.1 (b). See Figure 5.1 (a) for uniform weighting under the standard version
of the Colley method. In the linearly weighted Colley method, the weights
lie between 0 and 1 with the highest weight occurring on the last day of the
season. Games that occur on the same day are given the same weight, and
games on opening day are given no weight ($g(t) = 0$). Uncomfortable with
that choice? How about t_0 is 1 on opening day, which alters the weights on
all games and could result in a different ranking. Have another idea? This
could lead to a personalized bracket!

An important step in deriving (5.5) in the Colley method was the hovering
of the ratings about 1/2. An interested reader may wish to step through the
derivation of the Colley method to include a weighting function. We supply
only an intuitive sense of why one would expect such derivation to hold.
In the standard version of the Colley method, the outcome of a game increments one team's number of wins by 1 and the number of losses for the other
team by 1. In the weighted version of the method, the outcome of a game in-

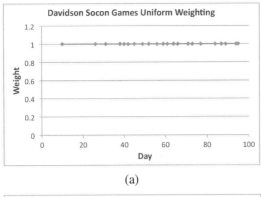

(a)

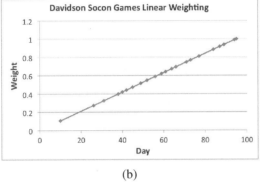

(b)

Figure 5.1. Weightings ((a) uniform, (b) linear) of games that Davidson played against Southern Conference opponents during the 2009 regular season. The scale of the y-axis differs on each graph.

crements the number of wins and losses of the associated teams by $g(t)$, so the important hovering about 1/2 still holds and the derivation that produced (5.5) is mirrored in the steps that produce an algorithm with a weighting function.

For a weighted method, the total games for team i, t_i, is altered. Instead of adding 1 to t_i for each game played we add $g(t)$. The alterations to the linear system $C\mathbf{r} = \mathbf{b}$ of the Colley method are minor. As before, the diagonal element of the Colley matrix c_{ii} is $2 + t_i$ but now t_i is this accumulation of games that includes weighting. The off-diagonal elements c_{ij} for $i \neq j$ equals $-n_{ij}$ where n_{ij} is the number of weighted games played between teams i and j. As with total games, n_{ij} is incremented by $g(t)$ for a game played between i and j at time t. Finally, b_i, which is an element of the vector $\mathbf{b}$ corresponding to team i, is $1 + \frac{1}{2}(w_i - l_i)$, where a game played on day t contributes $g(t)$ of a win to the winner and $g(t)$ of a loss to the loser.

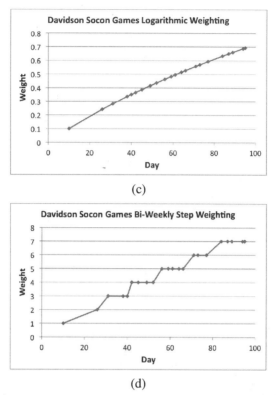

(c)

(d)

Figure 5.1. Weightings ((c) logarithmic, and (d) bi-weekly step functions) of games that Davidson played against Southern Conference opponents during the 2009 regular season. The scale of the y-axis differs on each graph.

5.4.2   Linear weighting in the Massey method

What alterations can incorporate weights into the Massey method? We begin with the initial Massey system $M_1\mathbf{r} = \mathbf{p}_1$ rather than the simplified system $M\mathbf{r} = \mathbf{p}$. Instead of applying least squares to $M_1\mathbf{r} = \mathbf{p}_1$, we use a weighted least squares method. Weighted least squares allows us to incorporate a measure of how much each game should count. If we have a low weighting for a game, say $r_i - r_j = 10$, then the algorithm will pay less attention to making that equation true when compared to a game with a higher weight. This works exactly as we would like in that if we weight more recent games heavily, the ratings that result from the least squares will give a rating that more closely reflects the outcome of the more recent games.

A weighted least squares method can be run by adding a weighting matrix W to the computation. Before we had M_1 which had each game on a different row. In the unweighted method, we calculated $M_1^T M_1 \mathbf{r} = M_1^T \mathbf{p}$ which led us to the final $M\mathbf{r} = \mathbf{p}$ system. We now add the weighting matrix, a square matrix whose diagonal entry w_{jj} corresponds to the weight of the jth game and has a value corresponding to the weighting function of the game. We then solve $M_1^T W M_1 \mathbf{r} = M_1^T W \mathbf{p}_1$.

5.4.3   Alternative weightings — when life isn't linear

If we are to weight games, how best to do it? While the linear model in (5.6) weights games at the end of the season higher than earlier games, we found this method was not as good as other choices. An alternative is logarithmic weighting

$$g(t) = \ln\left(1 + \frac{t}{t_f}\right), \tag{5.7}$$

which, without additional modifications, does not range from 0 to 1. As defined in (5.7), $g(t)$ produces weights as seen in Figure 5.1 (c) for games played by Davidson College. The function could also be defined so that it plateaus toward the end of the season and weights the more recent games nearly evenly.

Another option is to give the same weight to temporally similar games with a step function as is done with the following bi-weekly step function

$$g(t) = \lfloor (t - t_0)/14 + 1 \rfloor = \left\lfloor \frac{t}{14} + 1 \right\rfloor,$$

where $\lfloor x \rfloor$ represents the greatest integer less than or equal to x. This function divides the season into groups of 14 days and weights each group more than the previous two week period as seen in Figure 5.1 (d). This step function weights all games with integral values greater than or equal to 1. If a team wins a game on day 72 of the season, $g(72) = \lfloor \left(\frac{72}{14}\right) + 1 \rfloor = \lfloor 5.14 + 1 \rfloor = 6$. This win is equivalent to 6 wins in the unweighted (or uniformly weighted) formulation of the Colley method. We will see that the bi-weekly step function produced impressive results in predicting the results during March Madness in 2009.

Although our weighting methods depend on time, other non-temporal weightings are possible. Home court advantage can be accounted for if every away win is weighted higher than a neutral win, which is weighted higher than a home win. Another possibility is placing a small weight on games in which key players were injured. Weighting is an area of active research.

5.5   2009 Results

In this paper, we described two ranking methods, the Colley and Massey methods, and four weighting schemes (no or uniform weighting, linear, logarithmic, and step weighting), creating a total of $2 \cdot 4 = 8$ possible ratings that can be produced. Many more combinations exist, involving many other rating methods, weighting functions, and data inputs. The possibilities are limitless, which is part of the reason why in the past few years we have submitted several brackets to ESPN's Tournament Challenge. We enjoy seeing how brackets built from mathematical models compare both against each other and against non-mathematical submissions from other fans. The process for using a mathematical method to fill out a bracket is simple. First, we use all the data on games prior to the March Madness tournament to rate the teams. Then for each March Madness matchup in Round 1, we predict that the higher rated team will win. We do the same with the Round 2 matchups that we created, and so on for each round until we have completed a valid bracket. We follow the same process for each mathematical rating method that we submit to the ESPN Challenge.

For the 2009 March Madness tournament, we submitted several brackets. How did they fare? Out of the nearly 4.5 million brackets submitted, the Colley method with bi-weekly step weighting reached the 97th percentile, i.e., that bracket was better than 97% of the other brackets submitted. The previous year the bracket for the Massey method with logarithmic weighting was in the 99th percentile, putting it in the top 1000 of all 3.6 million brackets submitted. Table 5.5 provides a side-by-side comparison of our mathematical submissions alongside the non-mathematical submissions of some other famous fans.

While most of our mathematical submissions outperformed many of the non-mathematical submissions, one year the Colley method gave the best ESPN score yet the previous year the Massey method was the best. This is an issue with sports ranking. Although methods generally perform well from year to year, because of the inherent randomness of sports, it is difficult for a method to consistently take the title of the best mathematical method.

5.6   Concluding Remarks

Part of the popularity of March Madness relates to the difficulty of predicting the outcome of sports games. The linear models of this paper have performed quite well in producing brackets for March's basketball tournament. While the algorithms account for the strength of a team's opponents, there

Table 5.5. 2009 ESPN Tournament Challenge results (score and percentile) for the eight ranking methods and several non-mathematical submissions.

Mathematically produced brackets		
Algorithm	**Score**	**Percent**
Colley No Weighting	940	62nd
Colley Linear	940	65th
Colley Logarithmic	980	65th
Colley Bi-Weekly Step	1420	97th
Massey No Weighting	1300	88th
Massey Linear	1220	79th
Massey Logarithmic	1240	81st
Massey Bi-Weekly Step	1220	79th

Non-mathematically produced brackets		
Name	**Score**	**Percent**
President Barak Obama	1230	80th
Mike Greenberg (sports analyst)	1060	70th
Mike Golic (sports analyst)	800	43rd
Dwyane Wade (NBA star)	800	43rd

are countless possible weighting methods. What weighting would you use? Would you want to weight home versus away games, games in which key players were injured, or other statistics available about team performance? With a little thought and some luck you too could design a variation on these methods to fill out your own bracket and maybe even win next year's bracket challenge!

References

[1] W. N. Colley, Colley's Bias Free College Football Ranking Method: The Colley matrix explained, (2002) available at:
   `www.colleyrankings.com/method.html`.

[2] A. N. Langville and C. D. Meyer, *Recommendation Systems: Ranking, Clustering, Searching, and Visualizing Data*, Princeton University Press, Princeton, forthcoming.

[3] K. Massey, Statistical models applied to the rating of sports teams, undergraduate honors thesis, Bluefield College, 1997.

[4] T. Otterman, NCAA March Madness can cause lifetime gambling problems, *U.S. News & World Report*, March 17, 2009.

[5] Bowl Championship Series, available at: `www.bcsfootball.org`.

About the Authors

Timothy P. Chartier is an Associate Professor of Mathematics at Davidson College. He is a recipient of the Henry L. Alder Award for Distinguished Teaching by a Beginning College or University Mathematics Faculty Member from the Mathematical Association of America. As a researcher, Tim has worked with both the Lawrence Livermore and Los Alamos National Laboratories on the development and analysis of computational methods to increase the efficiency and robustness of numerical simulation on the lab's supercomputers, which are among the fastest in the world. Tim's research with and beyond the labs was recognized with an Alfred P. Sloan Research Fellowship. In his time apart from academia, Tim enjoys the performing arts, mountain biking, nature walks and hikes, and spending time with his wife and two children.

Erich Kreutzer is a student at Davidson College majoring in mathematics with a concentration in computer science. He is currently the president of the Bernard Society of Mathematics at Davidson. In the summer of 2007, Erich participated in the Google Summer of Code working on the Adium instant messenger application. His areas of interest include Mac OS X application development, sports ranking, data mining, and mathematical modeling. Outside of school Erich enjoys cycling, frisbee golf, and spending time with friends.

Amy N. Langville is an Associate Professor of Mathematics at the College of Charleston. Her award-winning book, *Google's PageRank and Beyond: The Science of Search Engine Rankings*, coauthored with Carl Meyer, explains and analyzes several popular methods for ranking webpages. Dr. Langville's research deals with ranking and clustering items using matrix decompositions. Her primary areas are numerical linear algebra, computational algorithms, and mathematical modeling and programming. Amy has consulted on six industrial projects with companies such as Fortune Interactive, Piffany, and the SAS Institute. She also regularly consults with law firms and is called to serve as a technical expert on patent infringement cases concerning webpage ranking systems. Off campus, Amy enjoys biking, basketball, and the outdoors.

Kathryn Pedings graduated from the College of Charleston in 2008 with a B.S. in Mathematics and a minor in Secondary Education. She is currently working towards a Master's in Mathematics at the College of Charleston and conducting research on sports ranking with Dr. Amy Langville while teaching full time at a local public high school.

CHAPTER 6

Down 4 with a Minute to Go

G. Edgar Parker

Abstract

According to the commentary on college basketball games, the prevailing wisdom concerning tactics associated with trying to win a game in which a team trails in the late stages, but is still close, is to "extend the game"; that is, to try to create as many more possessions as possible. Even when a team is down by only 3 points, announcers will advocate trying to score 2 points quickly rather than attempting to tie the game. In this paper, we analyze the situation in the title according to the points associated with the various outcomes from 2-point strategies and 3-point strategies, use the resulting model to suggest what the issues are in choosing a strategy for a particular team on the basis of its strengths or its opponent's weaknesses, and suggest how statistics might be gathered to inform the probabilities used in the model.

Play-by-play announcer : "The Hurrahs are down 81–77 with the ball coming out of this time-out and there are just 52 seconds left. I guess they'll shoot the three. Right, Billy?"

Expert analyst : No Jim, this is still a two-possession game. There is no need to panic; look for the easy two.

This conversation shows up time after time on both college and NBA telecasts. Apparently, if the announcing "experts" reflect the collective thinking of the coaches, this is the strategy that most coaches intend to play in this situation. In what follows plausible strategies associated with being down by four points at a time in the game when the opposition is in a position to, by exhausting the shot clock, limit a team to one additional possession are analyzed. The analysis examines the scenarios involved, and provides a theoretical model in which the strategy is dictated by the relative likelihoods of a team scoring two points or three points, or one point when trying for two or three, and the relative likelihood of the opponent scoring zero, one, or two points. The conclusion contains some questions raised by the analysis.

Suppose that the "easy two" strategy is an inferior strategy when compared to attempting a three-point shot[1] and being willing to attempt a second one if the opponent makes two free throws. Then why might most everybody employ the other strategy?[2] One possible reason is that when a team is down by four in the last minute of play, it is probably going to lose in regulation. Thus, when a team in that situation does win one or get to overtime, the game, and thus the strategy being used, is probably going to stand out in the minds of those who witness it. If most everybody uses the "easy two" strategy and such "stick-out" wins or overtime losses (a chance to win) are indeed likely to be memorable, then the rarity of such a win might reinforce the strategy rather than call it in question. To illustrate this, suppose that, in games of the type being considered, 90% of the coaches employ the "easy two" strategy and the other 10% employ a "shoot the three" strategy. Now suppose that there were 200 games where a team was down by four in the last minute and had the ball. Suppose further that the "easy two" strategy worked 5% of the time and the "shoot the three" strategy worked twice as often. There would be 9 games in which the "easy two" strategy worked; that means 18 head coaches (and probably 54 assistants) who are impressed with (or depressed by, in the case of the losing coaches) the results. Meanwhile, there would be only 2 games in which "shoot the three" worked; that would make only 4 head coaches and 12 assistants positively affected by the strategy. Empirically, the statistically better strategy attracts only 16 proponents while the inferior strategy attracts 72. These are not good odds in breaking down a stereotype.

Also important are factors that are actually under a coach's control. For instance, a coach cannot guarantee that her/his team will make a shot; she/he **can** be aware of the frequency with which the team has executed the desired strategies in practices or in past games. A coach cannot guarantee that the opponent will miss a foul shot; but a coach **can** be aware of the frequency with which the players on the court for the opponent have made foul shots in the past. Following are the strategies being compared.

---

[1]In case you haven't noticed the drift of the introduction, this, on the basis of the computation I have done with the model, is my position except in really exceptional situations.

[2]I would have guessed that, in the early days of the 3-point shot, that coaches would have thought of the model "3 + 3 = 6 (make two 3's); 6 − 2 (both free throws made) = 4 (the margin to be made up)", but it just must not have happened.

6.1   Shoot the 3

This strategy entails trying for a three-point field goal, and if a missed three-point field goal is rebounded by the shooting team, trying for another three-point field goal. Notice that there is also the possibility of making one, two, or four points. The shooter may be fouled and make only one or two of the free throws, a rebounder may be fouled and get only two free throws, or, in the case of an undisciplined opponent, the opponent may foul on a successful three-point shot and allow the shooting team to tie the game. The last of these outcomes is not allowed for in the model, although it adds minutely to the value of a three-point strategy.

6.2   Shoot the "easy" two

This strategy entails attempting to score a quick two-point field goal.[3] The rationale is that, since a defense is likely trying to prevent giving up three points and thus will guard the perimeter closely and is not likely to take a chance on fouling an inside shooter, that a field goal try from close range is more likely to be made than would ordinarily be the case.

Following the first possession, defensively, the strategy is typically to foul. Fouling allows the trailing team to regain possession of the ball **and** retain enough time[4] on the clock to execute a second strategic offensive possession. If there are fewer than 35 seconds to play, when playing a two-point strategy, there is no other option (unless the opponents are incompetent ball-handlers, a condition rarely encountered at higher levels of play), since time must be left to execute an offense and the team will trail by at least two points. In playing a three-point strategy, if there is time, the option to play defense for a shot clock is more attractive in the event three points have been scored, since even if the opponent scores two points in the possession, the possibility of tying or taking the lead in a final possession still exists.

The pertinent numbers associated with these strategies are the following:

- $T3$: the probability of scoring three points in a possession if the commitment is made to attempt a three-point shot as described in the three-point strategy.

---

[3]The dominant opinion voiced among broadcasters recently has been that coaches need "to extend the game". The Gambler's Road to Ruin Theorem, which says that the longer you play a losing strategy, the more certain you are to lose, indicates that the real issue is the one addressed in this paper.

[4]In 2004, Richard Miller investigated, statistically, the effect of time left on the shot clock on field goal percentage. He found trends in the data, but nothing statistically significant linking when a shot is taken during the shot clock's 35 seconds to field goal percentage.

- $t3$: the probability of scoring three points in a possession if the team is down by three points or fewer and commits to attempting a three-point shot.

- $D2$: the probability of scoring two points in a possession if the team is down by more than two points and commits to attempting a two-point shot.

- $d2$: the probability of scoring two points in a possession if the team is down by two points or fewer and commits to attempting a two-point shot.

- $T3_2$: the probability of scoring two points in a possession if the team commits to try for a three-point field goal when down by four points.

- $T3_1$: the probability of scoring one point in a possession if the team commits to try for a three-point field goal when down by four points.

- $D2_1$: the probability of scoring one point in a possession if the team commits to try for a two-point field goal when down by more than two points.

- $d2_1$: the probability of scoring one point in a possession if the team commits to try for a two-point field goal when down by two points or fewer.

- $D2_3$: the probability of scoring three points in a possession if the team commits to try for a two-point field goal when down by more than two points.

- $d2_3$: the probability of scoring three points in a possession if the team commits to try for a two-point field goal when down by two points or fewer.

- $X2$: the probability of the opponent scoring two points in a possession.

- $X1$: the probability of the opponent scoring one point in a possession.

- $X0$: the probability of the opponent not scoring.

Some relationships can be inferred or deduced between and among these numbers.

$D2 > d2 \geq T3 > t3 >> D2_3.$

Implicit in "easy two" is that the scoring percentage for an "easy two" should be higher than scoring a two when trailing by two, and scoring a two under "hard defense" should be no more difficult than scoring a three, even when the opponent is being careful not to foul a three-point shooter. Scoring three points when it will tie a game should be more difficult than scoring three points under the "don't foul" conditions. $D2_3$ should be negligible if the conditions for "easy two" are real.

$T3_1 > D2_1$ and both are tiny.

Implicit in the "easy two" is that the opponent dare not foul; a team up by four will not foul a three-point shooter. Thus whatever fouls occur are predominately rebound fouls. Rebounds are more likely to occur on three-point tries since they are more likely to be missed than "easy-two" attempts, and hence there should be a greater incidence of fouls on such attempts. The reluctance of the team with the lead to let the opponent score with the clock stopped should make both numbers very small, so omitting this factor is not likely to make much difference. For the same reasons, $T3_2$ should be tiny. $D2_3$ should be less than any of these since a foul on a two-point shot occurs under controlled defensive circumstances rather than loose-ball conditions. $d2_1$ is dependent on how often a team will be fouled when shooting a contested two-point field goal and miss one of the foul shots.

$X2 + X1 + X0 = 1$.

The opponent must score either zero, one, or two points under a "foul-them" strategy unless it rebounds a missed free throw after having made a free throw. Under such a possibility, the "easy-two" strategy must leave the trailing team in a position inferior to the one it was in before the two possessions, whereas a three-point strategy may have the team in the same position but with less time remaining. Thus, omitting the possibility of a three-point possession by the opponent, if it has an effect, skews the results in favor of an "easy-two" strategy. Given that a "foul-them" strategy is to be followed, $X2$, $X1$, and $X0$ can reasonably be computed from a single parameter, s, the percentage of free throws the opponent is expected to make.[5] The computation differs according to whether the opponent shoots in a 1-and-1 or 2-shot situation.

- For a 1-and-1 free throw situation, $X2 = s * s$, $X1 = s * (1 - s)$, and $X0 = 1 - s$.

---

[5]The assumption is that the team not shooting the free throws will rebound any miss. Some teams, to ensure that its players do not make a rebound foul or nullify a successful free throw by entering the three-second lane too soon, make this a certainty by not placing any players on the three-second lane.

- For a 2-shot free-throw situation $X2 = s * s$, $X1 = 2 * s * (1 - s)$, and $X0 = (1 - s) * (1 - s)$.

  When playing a three-point strategy and scoring three points in the initial possession, the opportunity exists to play defense for a shot clock. In this case, the numbers depend on offensive efficiency and $X1$ has a small probability relative to $X2$ and $X0$.

Given these parameters, probabilities can be computed for the outcomes after two of the trailing team's possessions associated with playing the different strategies by using the following models. Regarding the third possession (second offensive possession) in the sequence, the model is made on the assumption that a team down by three points in its (possibly) last possession, regardless of strategy, will make a three-point attempt.

Using an "easy-two" strategy: The probability of having a two-point lead is $D2_3 * X0 * d2_3$.

The probability of having a one-point lead is

$$D2_3 * X0 * d2 + D2 * X0 * d2_3 + D2_3 * X1 * d2_3.$$

The probability of being tied is

$$D2 * X0 * d2 + D2_1 * X0 * t3 + D2_3 * X0 * d2_1$$
$$+ D2 * X1 * t3 + D2_3 * X1 * d2 + D2_3 * X2 * t3.$$

Using a three-point strategy: The probability of having a two-point lead is $T3 * X0 * d2_3$.

The probability of having a one-point lead, using an aggressive strategy, is $T3 * X0 * d2 + T3_2 * X0 * t3 + T3 * X1 * t3.$[6]

The probability of having a one-point lead, using a conservative strategy, is $T3 * X0 * d2 + T3_2 * X0 * d2_3 + T3 * X1 * d2_3.$

The probability of being tied, using an aggressive strategy, is

$$T3*X0*d2_1+T3_2*X0*T3_2+T3_1*X0*t3+T3*X1*T3_2+T3*X2*t3.$$

The probability of being tied, using a conservative strategy, is

$$T3*X0*d2_1+T3_2*X0*d2+T3_1*X0*t3+T3*X1*d2+T3*X2*t3.$$

To interpret any of the calculations in the basketball context, reference the bullet list describing the pertinent numbers in the model. As an example,

---

[6]The defense fouling the shooter on a successful 3-point shot followed by a made free throw, followed by a successful two-point attempt in the next possession would also result in a two-point lead. But this is not accounted for in the model. As before, its omission skews the model in favor of an "easy-two" strategy.

consider "The probability of having a two-point lead is $D2_3 * X0 * d2_3$." This interprets as "The trailing team scores three points in a possession in which it committed to trying for a two-point field goal when down by more than two points ($D2_3$); the opposition scores no points in its possession ($X0$); the trailing team scores three points in a possession in which it commits to trying for a two-point field goal when down by two points or fewer ($d2_3$). ".

The assumptions behind these calculations are that all possessions constitute independent events and that if a team can take the lead with a two-point attempt, it will commit to running an offensive possession whose first priority is a two-point attempt. The numbers for an aggressive strategy are computed under the assumption that a team with the chance to take a lead with a three-point shot or to tie with a two-point shot will commit to running an offensive possession whose first priority is to create a three-point shot. The conservative strategy chooses to try for the tie. Also, to accommodate the strategy associated with playing defense for a shot clock rather than fouling, the values for $X2$, $X1$, and $X0$ associated with this strategy (that is, not consequent to deliberately fouling) may be used wherever the first factor is $T3$ or $D2_3$.

In order to effectively compare the three-point strategies with the two-point strategy, some way to measure the relative value of being two points ahead, one point ahead, and tied with the opponent must be devised. Here the factors used are probability of winning when tied in the last 35[7] or fewer seconds of a game and the opponent has the ball, when ahead by one point in the last 35 or fewer seconds of a game and the opponent has the ball; and when ahead by two points in the last 35 seconds or fewer of a game and the opponent has the ball. Letting $W2$ represent the probability of winning if a team has a two-point lead with 35 or fewer seconds remaining and the opponent has the ball, $W1$ represent the probability of winning if a team has a one-point lead with 35 or fewer seconds remaining and the opponent has the ball, and WE represent the probability of winning if a team is tied with 35 or fewer seconds remaining and the opponent has the ball, the models for the probabilities of winning are as follows.

Probability of winning playing an "easy-two" strategy :

$$W2 * D2_3 * X0 * d2_3 + W1 * (D2_3 * X0 * d2 + D2 * X0 * d2_3)$$
$$+ WE * (D2 * X0 * d2 + D2_1 * X0 * t3 + D2_3 * X0 * d2_1$$
$$+ D2 * X1 * t3 + D2_3 * X1 * d2 + D2_3 * X2 * t3).$$

---

[7]This would be 24 seconds for an NBA game.

Probability of winning playing an aggressive three-point strategy :

$$W2 * T3 * X0 * d2_3 + W1 * (T3 * X0 * d2 + T3_2 * X0 * t3$$
$$+ T3 * X1 * t3) + WE * (T3 * X0 * d2_1 + T3_2 * X0 * t3_2$$
$$+ T3_1 * X0 * t3 + T3 * X1 * t3_2 + T3 * X2 * t3).$$

Probability of winning playing a conservative three-point strategy :

$$W2 * T3 * X0 * d2_3$$
$$+ W1 * (T3 * X0 * d2 + T3_2 * X0 * d2_3 + T3 * X1 * d2_3)$$
$$+ WE * (T3 * X0 * d2_1 + T3_2 * X0 * d2$$
$$+ T3_1 * X0 * t3 + T3 * X1 * d2 + T3 * X2 * t3).$$

That $W2 > W1 > WE$ is a reasonable assumption; what the relative ratios actually are is not nearly as clear. Thus we allow for many possibilities to be accommodated when computing with the model.

Now it is time to test drive the model. Perhaps a good way to start is to find a "break-even" scenario. First, assume "competence" on the part of both offenses. Let the team that is leading be good free throw shooters by assigning a probability of .7 that any free throw will be made. This translates into $X2 = .49$, $X1 = .42$, and $X0 = .09$ in the model.[8] Let the team that is trailing be good on offense when trying to score two points by assigning $d2 = .55$[9] and be near-adequate three-point shooters, with the probability of .33 (1 out of 3) for making a three-point shot in a "don't foul" situation and .25 (1 out of 4) in a normal defense situation.[10] Set the "referee factor" data to 1 out of 50 for each outcome except "don't foul the three-point shooter" situations, to which I have assigned 0 probability ($T3_2$ and $T3_1$), and the situation of making a three-point play in a normal defense situation. I have chosen 1 out of 20 for this datum, since it happens during regular play more frequently than the other scenarios.[11] Finally, assign to the probability of scoring two points when trying for an "easy two" ($D2$), successively, the values .9, .8, .7, and .6 and assume that being tied is just as good as being

---

[8] For a 1-and-1 situation, these numbers are $X2 = .49$, $X1 = .21$, and $X0 = .3$.

[9] Note that this is not shooting percentage, but rather the percentage of possessions in which a team attempts a two-point shot as its first shot and scores exactly two points in the possession.

[10] 1 out of 3 is chosen since this has the same expected return as shooting 50% on two-point tries and 1 out of 4 is chosen because this has the same expected return as shooting 37.5%, which is less than 40%, the lower threshold that coaches are generally willing to tolerate and claim they actually have an offense. Neither ratio would be associated with having a team that is "good" at shooting three-pointers; 1 out of 4 would likely be associated with bad shooting.

[11] Note that, if these assignments show bias, it is towards a two-point strategy.

ahead by assigning each of WE, $W1$, and $W2$ the value 1. The model returns the following data:

Probability of being tied or ahead

	with	$D2 = .9$	$D2 = .8$	$D2 = .7$	$D2 = .6$
using					
"Easy two" strategy		.152156	.136256	.120356	.104456
Aggressive 3-point strategy		.096261	.096261	.096261	.096261
Conservative 3-point strategy		.096261	.096261	.096261	.096261

Notice that how easy an "easy two" is is a critical issue. If an "easy two" is a virtual surety, the probability of being tied or ahead is half again as large as that associated with a three-point strategy where the shooters are marginally accurate. **But**, if "easy two" means 9% better than an already efficient offense, a two-point strategy is only slightly better, judged on being even or ahead, with a three-point strategy where the shooters are only marginally accurate.

On the other hand, being ahead gives a better chance of winning than being tied, so perhaps ignoring the probability of winning associated with being ahead as opposed to being tied is not a very practical idea. Use WE, the conditional probability of winning given that the game is tied and the opponent of the formerly trailing team has the ball with less than a shot clock remaining or time has expired, as the basic statistic. If the formerly trailing team takes a one-point lead, depending on the time remaining, the team that was previously leading either has time to score or it doesn't. If the previously leading team scores two or three points, it wins (as it would if the game were tied); if it scores one point, overtime is played (if the game were tied, this would result in a win for the scoring team); or if it does not score, the formerly trailing team wins. Thus wins for the formerly trailing team result in all scenarios in which a tie results in a win plus those in which the other team fails to score. So $W1 > WE$ and $W1 - WE$ must include the value of getting to overtime should the formerly leading team score one point and the value of winning when the formerly leading team fails to score. Similarly $W2 > W1$, since the formerly leading team must score two points to force overtime and three points to win, and scoring one point, a part of the gain in $W1$ over WE, now results in a loss for the formerly leading team. $W2 - W1$ must include the value of getting to overtime associated with the formerly leading team scoring two points and value for a win for the formerly trailing team resulting from the formerly leading team scoring only one point.

All of WE, $W1$, and $W2$ depend on the probability of the formerly leading team scoring in a final possession after the score has been tied or it has
relinquished the lead. It is reasonable to set this probability lower than the
probability of scoring in a full shot clock possession since there is the possibility of there being only a few seconds left and, in a college or high school
game,[12] the team must travel the length of the court. Let V represent the
probability of the formerly trailing team winning, given that an overtime is
played, p represent the probability of the formerly leading team scoring in
a final possession, and p' represent the probability of it scoring exactly one
point if it scores two points or one point. Then

$$WE = (1 + -p) * V,$$
$$W1 = (1 + -p) + (p * p' * V),$$

and

$$W2 = (1 + -p) + (p * p') + \big((p + -p') * V\big).$$

Suppose that winning in overtime is a break-even proposition, that the
formerly leading team is expected to score one or two points 20% of the
time in its final possession, and that 5% of scoring possessions resulting in
no more than two points result in a single point. Then $V = .5$, $WE = .4$,
$W1 = .805$, and $W2 = .8525$.

Using the same data for all other factors as before, but with the weights
giving proportional value for the likelihoods of winning suggested above,
the model returns the following data:

Probability of winning

using	with $D2 = .9$	$D2 = .8$	$D2 = .7$	$D2 = .6$
"Easy-two" strategy	.063114	.056572	.050030	.043488
Aggressive 3-point strategy	.059825	.059825	.059825	.059825
Conservative 3-point strategy	.066894	.066894	.066894	.066894

*Against a good free-throw shooting team, even with an extremely high proba-
bility of scoring the "easy two" and only marginally good three-point shooting, a conservative three-point strategy is superior to an "easy-two" strategy
and an aggressive three-point strategy passes from break-even to superior
between .9 and .8, and at "easy-two" levels 10% above normal offensive
production, both are 40% to 50% better.*

---

[12]NBA rules allow a team to take the ball at half court if it chooses to call time out.

But what about a poor free-throw shooting team? The following data is based on the team with the lead expected to make only half of its free throws. This translates, for the double bonus situation, into $X0 = .25$, $X1 = .5$, and $X2 = .25$.[13] Using the other input data as before, the model returns the following data:

Probability of being tied or ahead/Probability of winning

	with	$D2 = .9$	$D2 = .8$	$D2 = .7$	$D2 = .6$
using					
"Easy-two" strategy		.255910	.231600	.204100	.176600
		/.109626	/.098119	/.086613	/.0751069
Aggressive 3-point strategy		.116325/.083480			
Conservative 3-point strategy		.116325/.091895			

Here, the shift from a two-point strategy being the superior strategy to a conservative three-point strategy being superior comes between 80% and 70% efficiency defining the "easy two" and for an aggressive three-point strategy between 70% and 60%.

But what if, against a poor free-throw shooting team, the trailing team is actually good at shooting three-point field goals? Assign .4 to $T3$ and .33 to $t3$;[14] that is, quantify an expectation of making 2 of 5 in a "don't foul" situation and 1 of 3 in a "normal defense" situation.

The model returns the following data:

Probability of being tied or ahead/Probability of winning

	with	$D2 = .9$	$D2 = .8$	$D2 = .7$	$D2 = .6$
using					
"Easy-two" strategy		.295900	.264400	.232900	.201400
		/.124346	/.11239	/.098133	/.085027
Aggressive 3-point strategy		.165000/.117268			
Conservative 3-point strategy		.165000/.114588			

This chart differs from the others in that an aggressive three-point strategy is superior to a conservative three-point strategy. In all of the above scenarios, both three-point strategies are comparable to the two-point and, unless an "easy two" is **really** easy relative to the normal chances of scoring, superior. Also keep in mind that **lowering the "normally- defended two" efficiency OR raising the "don't foul the 3-point shooter" efficiency** *widens the gap.*

---

[13] For a 1-and-1 situation, these numbers are $X2 = .25$, $X1 = .25$, and $X0 = .5$.

[14] These are good numbers, but not outlandishly high.

So there are the model and some suggestive data. Rev up the computer and go to work! Create plausible data for the factors in the model. Investigate how BAD the leading team's free throw shooters have to be relative to how bad the trailing team's three-point shooters have to be in order to make the "easy-two" strategy the superior strategy. Or better yet, make charts that the coaches can use that show "break-even" values for the strategies and will allow the coach to size up, on the spot, which strategy is likely to be best on a given night.

But wouldn't it be nice to know what these input probabilities typically are and to have observational data that comment on how real things like the "easy two" are? I suggest that statistically reliable answers to the questions posed below could carry the model from "intriguing and highly suggestive" to "here's what a coach *should* do".

Question # 1   For some statistically significant sample of games, what is the ratio of number of games in which a team was behind by 4 and had the ball in the final minute and won or forced overtime to the number of games in which a team was behind by 4 and had the ball in the final minute?

Prediction   This ratio will be a small number. Interesting related question: Does this ratio change appreciably among levels of play, i.e. high school boys, high school girls, college men, college women, Division I, Division III, NBA? Important side data — record the outcomes of the possessions starting with down four with the ball and less than a minute to play.

Question #2   What do basketball coaches do when they are down by four points, have the ball and there is less than a minute left to play?

What are their strategies on this possession? What are their rationales for what they do?

Question #3   How do the shooting percentages on two-point tries for teams behind by three or more late in a game compare to the percentages for two-point field goals in the rest of the game preceding the situation?

I would be amazed if they were appreciably different. I suspect that the "easy two" is a myth, but its existence is the only basis I can imagine for using a two-point strategy. The data offered in the charts indicates that its substance is an issue of vital importance in selecting a strategy at the end of a game.

Question #4   Is there any difference in the likelihood of scoring when a team trails by two points as compared to being behind by one point?

Question #5   What are the numbers, as observed in actual games, associated with the factors in the model?

If the rationale on which the easy-two strategy is based is correct, $D2$ should be higher than $d2$. **Note that neither of these is field goal percentage; each is a ratio of times a team scored exactly 2 points during a possession in which the initial shot is a 2-point try to the number of possessions in which the initial shot is a 2-point try.** For $D2$, the possessions need to be counted in the pertinent situation, trailing by four points with less than a minute to play; for $d2$, they need to be counted when trailing by two points late in a game. There is no reason to believe that $T3_1$ and $D2_1$ are appreciably different since 3-point attempts are seldom fouled, and shouldn't ever be fouled in this situation. Moreover, implicit in the "easy-two" rationale is that the opposition will not risk fouling, or as the commentator's say: "let a team score while the clock is not running". Is it safe to take $X2$, $X1$, and $X0$ to be the ratios for the rest of the game if you are playing defense for an entire shot clock? Or is there something real about clutch performance (offensively or defensively)? What are the typical relative sizes of $X2$, $X1$, and $X0$ when a team is employing a "foul-the-opponent" strategy? These statistics for $X2$, $X1$, and $X0$, since they are so closely tied to the ability to shoot free throws, should be the most convenient for coaches to gather if they were to choose to use the model.

Question #6   What are reasonable numbers for $W2$, $W1$, and WE? Data gleaned from the model in the charts without introducing $W2$, $W1$, and WE strongly suggest that three-point strategies give comparable chances of being tied or ahead when $D2$ is within 10% of $d2$. But, since being ahead after three more possessions is a possible outcome in a three-point strategy whereas the best outcome, **except in extremely unlikely circumstances,** from a two-point strategy after three more possessions is a tie, introducing $W2$, $W1$, and WE will likely show how much more powerful a three-point strategy is, even without the assumption of proficient three-point shooting.

I'm guessing, if the questions were answered consequent to reliable data, that, scientifically at least, the issue of the superior strategy would be situationally settled. That is, unless the leading team is known to shoot free throws abysmally and such shooters are on the court to be fouled, or unless there is no one on the trailing team who can make a three-point field goal,[15] "extending the game" would be flying in the face of wisdom. Until that time,

---

[15]I once heard Rick Majerus say in an interview: "I've got lots of players who can shoot the three; I just don't have anybody who can make one."

as a basketball TV spectator, my experience will remain the same. Every time an "expert" tells me, "No need to panic and shoot a three.", I will yell back at the set, "You're right, they need to shoot **two** 3's!".

References

[1] Richard Miller, "Mathematical Analysis of Sport" BIS Capstone Project, James Madison University, December, 2004.

About the Author

G. Edgar Parker received his BS degree in Mathematics with a minor in Philosophy and Religion from Guilford College in 1969 and his PhD in mathematics from Emory University in 1977. In a teaching career that now spans 39 years, Ed has taught in high school, junior college, an open-admissions university, and selective universities, and remains committed to Inquiry-Based Learning. Although the bulk of his research has been done on differential equations, sports (in particular baseball) are his great recreational passion, and he is often led to think about sports due to listening to or reading remarks from the media or coaches as they analyze games he has seen. While an undergraduate, Ed was on the basketball team his freshman year and played varsity baseball for four years. He coached junior varsity basketball and varsity baseball during his four years (1969–1973) at Bayside High School in Virginia Beach, Virginia and has actively promoted youth sports participation in all of the places he has lived.

CHAPTER 7

Jump Shot Mathematics

Howard Penn

Abstract

In this paper we examine variations of standard calculus problems in the context of shooting a basketball jump shot. We believe that many students will find this more interesting than the use usual manner in which such problems are presented in text-books.

7.1 Angle of elevation 60 degrees

Suppose a basketball player takes a 15 foot jump shot, releasing the ball from a height of 10 feet and an angle of elevation of 60 degrees. What is the initial speed V_0 needed for the shot to go in?

If we neglect air resistance, this is a typical ballistic motion problem. The equations are [1]

$$x(t) = V_0 \cos(\theta)\, t, \quad y(t) = -\frac{gt^2}{2} + V_0 \sin(\theta)\, t + h_0.$$

For this problem we have $g = 32$ ft/sec^2 and $\theta = 60°$. Since the ball is released from the height of the basket, we can take h_0 to be zero.

The range is given by

$$d = \frac{V_0^2 \sin(2\theta)}{g}.$$

If we set $d = 15$ feet and solve for V_0 we get,

$$V_0 = \sqrt{\frac{15 * 32}{\sin(120°)}} \approx 23.5426 \text{ ft/sec.}$$

Suppose that later in the game, the player takes another jump shot from the same position but, maybe because she has tired, takes the shot with an initial angle of elevation of 30 degrees. What is the initial velocity needed this time?

Since the angles are complementary, the initial speed will be the same:

$$V_0 = \sqrt{\frac{15 * 32}{\sin(60°)}} \approx 23.5426 \text{ ft/sec.}$$

Figure 7.1 shows the path of the two shots.

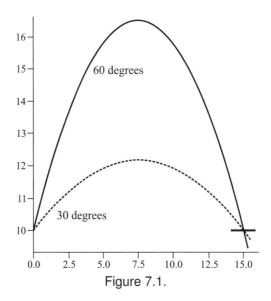

Figure 7.1.

In Figure 7.1, the lower curve is closer to the front rim of the basket than the upper curve. Basketballs are not points. According to Wikianswers.com [2], the rim of a basket has a diameter of 18 inches. A woman's basketball has a diameter of approximately 9 inches and a man's basketball has a diameter of about 9.4 inches. For the rest of this paper, we will assume that the shooter is a woman basketball player.

We are interested in how close the center of the basketball comes to the front rim or back rim. Put another way, does the shot result in nothing but net?

This is a variation of an optimization problem seen in most calculus books (see [1]), namely, finding the minimum distance from a point not on a curve to the curve. We believe that students do not find these problems interesting, but some students will be interested in the answer to our basketball problem. For the shot at an angle of 60 degrees, the equations of motion are

$$x(t) = V_0 \cos(60°)t,$$
$$y(t) = -16t^2 + V_0 \sin(60°)t.$$

The distance from the front of the rim is given by the minimum of

$$Df = \sqrt{(x(t)14.25)^2 + (y(t))^2}.$$

For Db, the distance from the back of the rim, 14.25 is replaced by 15.75. The results are $Df_{min} = 7.77$ in. and $Db_{min} = 7.82$ in.

The shot easily goes through the basket touching nothing but net using either the man's or the woman's basketball.

7.2 Angle of elevation 30 degrees

With an initial angle of 30 degrees, the equations are the same except that 30 degrees replaces 60. We will leave it as an exercise to verify that $Df_{min} = 4.37$ in. and $Db_{min} = 4.62$ in.

Since the radius of the ball exceeds 4.37 in., it hits the front of the rim, but would clear the back of the rim with room to spare. Figure 7.2 shows the paths near the basket with the balls and basket superimposed.

Perhaps we can make the shot if the center of the ball goes through a point a little closer to the back of the rim. After experimenting with changing the

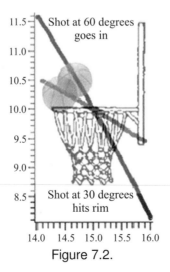

Figure 7.2.

initial velocity, we found that with an initial speed is 23.56 ft/sec we obtain $Df_{min} = 4.49$ in and $Db_{min} = 4.48$ in. Since the radius of the basketball is 4.5 in, the shot grazes both the front and back rims. Thus it is not possible to get nothing but net shooting at an angle of elevation of 30 degrees from 15 feet if the shot is released from a height of 10 feet.

This raises the question: how close to 30 degrees can we get?

Solving the equations for a shot through the center of the basket with an initial angle of 31 degrees yields

$$V_0 = 23.31 \text{ ft/sec,}$$
$$Df_{min} = 4.505 \text{ in,}$$
$$Db_{min} = 4.756 \text{ in,}$$

so we can conclude that the ball clears the rim and the shot is good.

7.3   Varying the distance

What if the shot is taken from a distance other than 15 feet? When we first considered this problem, it seemed that we would have to solve the equations with an additional variable, the distance from the basket. We can make the problem simpler if we imagine a defender trying to block the shot and having the tips of her fingers at a height of 10 feet and a distance of 9 inches from the center of the ball at release. Since the release point and basket are the same height, the minimum distance of the ball from the front of the rim is exactly the same as the minimum distance of the ball from our imaginary defender.

If we eliminate the parameter by solving for $t = x/(V_0\sqrt{3}/2)$ and substitute into $y(t)$, we get

$$y = \frac{-16x^2}{V_0^2} + \frac{\sqrt{3}x}{3}.$$

This is the equation of a parabola opening downward, so the path is concave down and lies below its tangent line at $x = 0$. As we increase the distance from the basket and hence the initial velocity, the path approaches that tangent line. The equation of the line tangent to the curve at $x = 0$ is

$$y = \frac{\sqrt{3}x}{3}$$

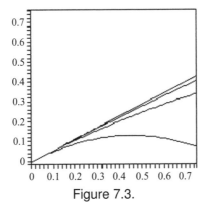

Figure 7.3.

and the minimum distance from the defender's hand to that line is

$$Df_{\min} = 4.5 \text{ in.}$$

Trigonometry makes the minimum distance to the defender's hand obvious since it is the length of the side opposite a 30 degree angle in a right triangle, and half the length of the hypotenuse, the distance from the center of the ball at release to the defender's finger tips, 9 in. Since the actual path of the ball lies below this tangent line, the minimum distance to the defender's fingers must be less than 4.5 in and the shot is blocked.

Figure 7.3 shows the paths $(0 < x < 0.75)$ of the center of basketballs shot with initial velocities of 5, 10, 20 ft/sec and the line tangents to them at $x = 0$. The tip of the defender's fingers is the lower right corner of the graph.

If we view shots taken at different initial velocities at an angle of 30 degrees as the basketball passes through the center of the basket, we get the mirror image of Figure 7.3. Thus any shot at an initial angle of elevation of 30 degrees that goes through the center of the basket will have a minimum distance from the front rim less than the 4.5 in radius of the ball and will therefore touch the front rim.

7.4 Varying the height

Another variation is to consider a shot from 15 ft at an initial height different than 10 ft. Since the path is a parabola, the closer we are to the vertex, the smaller the angle of elevation. If a shot is taken from a height of less than 10 ft, the angle at the basket will be less than 30 degrees and thus can't clear both rims. If the shot is taken from a height greater than 10 ft, the angle at the basket will be more than 30 degrees and it will be possible to miss both the front and back rims. For example, if the jump shot is from a height of 9 ft, the initial speed will be approximately 25.03 ft/sec, the angle of elevation at the basket is approximately -23.94 degrees, and $df_{min} \approx 3.50$ in. On the other hand, if the shot is taken with an initial height of 11 ft, the initial speed will be approximately 22.29 ft/sec, the angle of elevation at the basket is approximately 35.40 degrees, and $df_{min} \approx 5.10$ in. Since the basketball has a radius of 4.5 in, the first shot hits the rim while the second one clears the rim and scores.

What does this say about a shot at an initial angle of elevation of 30 degrees? First, we see that it is not possible to make the shot unless the initial height is more than 10 feet. Second, the shot at the low angle is more likely to be blocked and third, a higher angle of elevation means that the ball will be in the air longer so you and your team will have more time to get into position for the rebound.

After all this there *is* a way to make a shot from a height of 10 feet with an initial angle of elevation of 30 degrees. The ball must be shot with the center of the basketball at a distance from the center of the rim less than 4.5 inches! Of course, this requires the player to be under the basket and shot from inside it.

References

[1] J. Stewart, Calculus, *Early Transcendentals 6th ed.,* Thompson, Brooks/Cole. Belmont, CA 2008

[2] Wikianswers available at wikianswers.com

About the Author

Howard Penn received his B.A. in Mathematics from Indiana University in 1968 and his Masters and PhD in Mathematics from the University of Michigan in 1969 and 1973. Since receiving his final degree, he has taught at the United States Naval Academy where he now holds the rank of Professor. He is one of the pioneers is the use of computer graphics in the teaching of mathematics, having written the program, MPP, which was widely used internationally. He is also interested in finding applications of the mathematics to areas that many students will find interesting. Outside of mathematics, he is well known amateur photographer with special interests in wildlife, landscapes and abstract photography. His work may be viewed at flickr.com/photos/howardpennphoto.

Part III

Football

CHAPTER 8

How Deep Is Your Playbook?

Tricia Muldoon Brown and Eric B. Kahn

Abstract

An American football season lasts a mere sixteen games played over seventeen weeks. This compact season leads to, fairly or unfairly, intense scrutiny of every player's performance and each coach's decision. Our goal in this paper is to determine a measure of complexity for the decision of choosing a defensive alignment on any given down. There are only four standard defensive formations, defined generally by the personnel on the field, but how a coach physically situates the players on the field can emphasize widely different defensive strengths and weaknesses. To describe the number of ways a coach achieves this goal, we utilize the notion of equivalence classes from abstract algebra to define classifications of defensive formations. Enumerative combinatorics is then necessary to count the number of fundamentally different defensive alignments through the application of binomial coefficients. Descriptions of the rules for the game and diagrams of different defensive alignments make this paper accessible to even the novice, or non-fan.

8.1 Introduction

Many people have fond memories of participating in team sports in elementary school, high school, and even college. We recall the thrill of a close game and the joy or sadness after a win or loss. As we age, less of our time is devoted to playing sports as other responsibilities take priority and our bodies simply are not capable of competing at the same level. Our role in sport transforms from being an active participant to being a spectator. Many of us still go to the gym, play a game of racquetball, or pick up the old golf clubs on a regular basis, but rarely do these friendly competitions end in the

93

glory of a plastic trophy so coveted in our past years. Now as we focus our attention to watching our favorite professional, college, or club teams, our competitive edge discovers a different outlet: critiquing athletes and coaches who are still involved in the games we love. We consider one such game.

One cold Sunday during a snowy December, residents of Pennsylvania and Massachusetts settled down in front of their televisions to watch a classic sport, American football. The Pittsburgh Steelers are hosting one of their chief rivals, the New England Patriots, at Heinz field and the game is almost over. It has been a close game, and the Steelers have the ball with time for one last play. The populations of two states are waiting anxiously to see how the two teams will line up, when one insightful and mathematically inclined fan decides to try to calculate the magnitude of complexity of Patriots coach Bill Belichick's decision to pick a defensive alignment. Before she can begin her calculations and before we can get back to the final play of the game, we describe the background needed to understand this question.

8.2   The Game of Football and Mathematics

The game of football is a competition between two teams on a playing field that is 100 yards long and 53.3 yards wide. In the National Football League, each team is composed of 53 players of whom exactly 11 are on the field at any given time. The goal of the team on offense is to traverse the length of the field and reach the far end; the goal of the defense is to prevent their opponents from completing this task. To fix our orientation in space, we will view the field as being longer in the vertical direction and shorter in the horizontal. Defensive formations will always align facing down the field. The right and left sides of a formation will be with respect to the orientation of the reader.

There are typically four different positions or types of players in a defensive formation: linemen, linebackers, safeties, and cornerbacks. We will assume there is no distinction between different players of the same type. Different combinations of players on the field at one time determine a formation. The four basic formations used by a defense during a football game are: the 3-4 which is made up of 3 linemen and 4 linebackers, the 4-3 which is made up of 4 linemen and 3 linebackers, the nickel which is made up of 4 linemen, 2 linebackers, 3 corners, and the dime which is made up of 4 cornerbacks, 2 safeties, and at least one linebacker. Diagrams displaying the standard alignment of each formation are in Figures 8.1–8.4.

To organize vertically how far from the line of scrimmage players stand, the field is divided into tiers or levels stretching its entire width. The levels from front to back are called: defensive line, mid, nickel, and deep. To organize the field horizontally, we consider the middle 23.5 yards to be the interior of the field and the 15 yards on either side of the interior to be the exterior, dividing the field into three sections. Restrictions determining where different types of players may align will be determined after we discuss the necessary mathematics. To answer our question we make use of two branches of mathematics, enumerative combinatorics and abstract algebra.

One of the goals of enumerative combinatorics is to count sets of objects satisfying certain conditions. That is, we consider the number of ways a given pattern or formation can be created. We will need to count *combinations*, unordered subsets of a given set. The number of ways to choose a combination of k elements from a set of n distinct elements is given by the binomial coefficient

$$\binom{n}{k} = \frac{n!}{k!(n-k)!}.$$

One of the goals of abstract algebra is to describe formally the structure and relationships between objects in a set or between the sets themselves. The notion of an equivalence relation plays an important role in our problem. An *equivalence relation* $\mathcal{R}$ on a set S is a subset of all pairs in $S \times S$ that is reflexive, symmetric, transitive. A standard example of an equivalence relation on the set of real numbers $\mathbb{R}$ is $x\mathcal{R}y$ if $|x| = |y|$.

We will use a similar type of relation while describing defensive alignments. Two defensive alignments will be termed *mirror equivalent*, or simply *equivalent*, if at each level the order of the players from right to left is reversed. For this to be an equivalence relation, we also require a defensive alignment to be mirror equivalent to itself. The goal is to count the number of non-equivalent defensive alignments, thus combining combinatorics and algebra to describe a game many Americans watch every Sunday during the fall and winter months. However, we want the defensive alignments that are counted to represent actual situations. The following is a list of restrictions we will use in creating different alignments for a defense. The list is by no means exhaustive but it allows for most defensive alignments found during a football game.

1. Linemen can line up only on the defensive line level.

2. There must be at least two linemen on the field.

3. At most two linebackers can line up at the defensive line level.

4. All other linebackers must line up at the mid level.

5. Corners can line up at any level.

6. Safeties can line up at the deep or mid level.

7. There must be exactly two safeties on the field with at least one at the deep level.

8. Only corners will line up in the exterior regions of the field outside the linebackers and safeties.

9. Two linebackers cannot stand side by side on the defensive line level.

8.3   Counting the Formations

We will first determine which players are fixed, e.g., the linemen are always on the line and one safety is always deep. We then insert the remaining types of players into the formation, first the linebackers, then the other safety, and then the corners. At each stage we consider whether a formation is *self-symmetric*, denoted by *ss*, that is, the alignment is mirror equivalent only to itself. When inserting a player into a self-symmetric formation, we count equivalent placements or reflections only once. Once an alignment is not self-symmetric we count both placements. This is because in a mirror equivalence the size of the equivalence class may only be one or two. If an alignment is not self-symmetric, it has already been paired with its reflection. It is a representative of an equivalence class of size two. Thus, each place a new player can be added creates a distinct formation. We place the players in the interior of the field in this fashion, keeping count of which alignments are self-symmetric. Corners are the only players who line up on the exterior sides of the fields. It will be left to count the number of ways the corners can line up with respect to the mirror equivalence and without. We multiply the former by the number of self-symmetric formations, that is, representatives of equivalence classes of size 1, and multiply the latter by the number of formations which are not self-symmetric. We consider each of the four types of formations.

The 3-4 Defense

The 3-4 defense is designed to be flexible enough to be effective against both running and passing plays. Figure 8.1 has an example of a 3-4 formation. Because of the restrictions, the three linemen, two linebackers, and one safety are fixed at specific levels. Next, there are three scenarios for placing the other two linebackers.

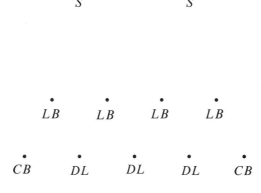

Figure 8.1. A standard 3-4 formation

1. Both on the line. This can be done in three ways:

   (a) both outside all linemen (ss), one way

   (b) one between two linemen and one outside all linemen in two ways, that is, either one or two linemen between the two linebackers on the line

   (c) both between two linemen (ss), one way.

2. Both on mid level (ss) one way.

3. One on the defensive line and one at the mid level. There are two spots for the one on the line:

   (a) inside in one way

   (b) outside in one way.

Now we place the remaining safety. For each formation the safety can be placed at the deep level in exactly one way. That is, we have three self-symmetric formations and four formations that are not self-symmetric where both safeties are deep. We determine the formations for the safety on the mid level.

1. Parts (a) and (c) are self-symmetric, so by mirror symmetry, the mid level safety can be placed in each formation in two ways, between the two line backers (ss) and outside the line backers. Neither of the formations in part (b) is self-symmetric, so we count each position for the

safety. He can stand in three different ways, on the outside on either side or in between the two line backers.

2. This formation is self-symmetric. We have three ways to place the last safety; in the center of the mid level with two line backers on either side (ss), between two linebackers with one on one side and three on the other, and outside the linebackers.

3. Neither part (a) nor part (b) is self-symmetric. Therefore, the safety can be placed on mid level in four ways, between the linebackers two ways and outside the linebackers two ways, for each formation.

So, we have $1 \cdot 2 + 1 \cdot 1 = 3$ self-symmetric formations with one safety on the mid level and $1 \cdot 2 + 3 \cdot 2 + 2 \cdot 1 + 4 \cdot 2 = 18$ formations where the safety is on the mid level and which are not self-symmetric. Including the cases where both safeties are deep we have six self-symmetric formations and 22 alignments which are not self-symmetric. See figures 8.5, 8.6, and 8.7 for the progression of aligning the interior section of the 1b. alignment.

It is left to position the two corner backs. We have two cases.

1. In the first, both corners stand on the same side and same level. This can be done in eight ways. Applying the mirror equivalence, each of the eight formations where the corners are standing on the same line and same side are symmetric to exactly one other. We have four different formations respecting mirror equivalence.

2. In the second case, each corner stands on a different side or level. We have $\binom{8}{2} = 28$ possible places to line up the corners. With mirror symmetry there are four formations of the 28 which are self-symmetric, namely the ones where the corners line up on the same line. That leaves 12 placements that have been counted twice, that is, 4+12=16 total alignments with respect to mirror equivalence.

That gives 8+28=36 possible formations without regard to symmetry which we apply to the formations of linemen, linebackers, and safeties that are not self-symmetric. There are 4+16= 20 possible formations with regard to the mirror equivalence that we apply to the self-symmetric formations. Thus we have

$$20 \cdot 6 + 36 \cdot 22 = 912$$

possible 3-4 defensive formations.

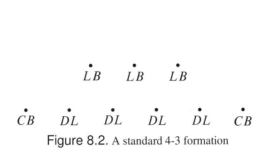

Figure 8.2. A standard 4-3 formation

The 4-3 Defense

The 4-3 defense is common among NFL teams and is designed to defend against offensive running plays. The personnel on the field are always four linemen, three linebackers, two corners, and two safeties. See Figure 8.2. From our restrictions, the four linemen must line up on the defensive line level and one linebacker must line up at the mid level. The other two linebackers have the freedom to line up at the mid or defensive line levels.

1. If both linebackers are placed at the mid level then there is only 1 possible formation (ss).

2. If one linebacker is at the mid level and one is at the defensive line level there are 3 formations:

   (a) outside all linemen one way

   (b) in between the linemen with one on a side and three on the other side one way

   (c) in between the linemen with two on each side 1 way (ss).

3. If both linebackers are at the defensive line level then there are three formations:

   (a) both outside all linemen one way (ss)

   (b) one outside all linemen and one in between two linemen three ways

   (c) both in between linemen with exactly two linemen between the linebackers one way (ss)

   (d) one with two linemen on either side and one with a single lineman on one side and three on the other one way.

Next we place the remaining safety. In each case, the remaining safety may be placed at the deep level in exactly one way and this will respect the self-symmetries, that is four self-symmetric formations and six that are not self-symmetric. Next we must count the number of formations with the safety at the mid level.

1. The mid level safety can be placed outside all linebackers or in between two linebackers for a total of two ways, neither of which is self-symmetric.

2. Parts (a) and (b) are not self-symmetric. The mid level safety can be placed in between the linebackers or outside the two linebackers in three ways resulting in six new formations which are not self-symmetric. Part (c) is self-symmetric, so the safety can be placed on the mid level in two ways, only one of which is still self-symmetric.

3. Parts (a) and (c) are self-symmetric, so the mid level safety is placed on either side of the lone linebacker in one way for each formation. Parts (b) and (d) are not self-symmetric, hence the safety can stand on either side of the linebacker in two ways for each scenario. None of the new alignments are self-symmetric.

A defense will be self-symmetric if and only if all levels of the defense are self-symmetric, meaning we have four self-symmetric formations with both safeties deep and exactly one self-symmetric formation with one safety at the mid level. For the non-self-symmetric formations we have six formations with both safeties deep and with a safety at the mid level we have:

$$1 \cdot 2 + 2 \cdot 3 + 1 \cdot 1 + 2 \cdot 1 + 4 \cdot 2 = 19$$

formations. Thus we have five self-symmetric formations and 25 non-self-symmetric formations.

Lastly we must place the remaining two cornerbacks. In the 4-3 defense, there are the same number of cornerbacks as in the 3-4 defense and each has the same options of where to line up in the 3-4 defense. Thus, counting the alignments of cornerbacks in the 4-3 will be identical to that of the 3-4. So there are 36 possible alignments without regard to the mirror symmetry that we apply to the non-self-symmetric formations of linemen, linebackers, and safeties and 20 possible formations that we apply to the self-symmetric formations. This gives a total number of distinct defensive alignments in the 4-3 defense to be

$$36 \cdot 25 + 20 \cdot 5 = 1,000.$$

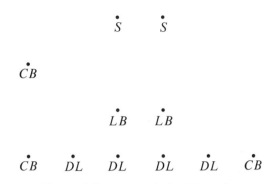

Figure 8.3. A standard Nickel formation

The Nickel Defense

The nickel defense is an altered version of the 4-3, primarily to protect against offensive passing plays. An example can be found in Figure 8.3. From our restrictions, the four linemen must line up on the defensive line level and both linebackers have the freedom to line up at the mid or defensive line levels. In terms of counting the alignments for linemen and linebackers, this is equivalent to the 4-3 formation as the extra linebacker in the 4-3 is fixed and thus has no effect on the counting problem. We omit the list of possible formations and instead will reference the linebacker possible alignments from the 4-3 formation while counting the rest of the nickel defense alignments.

Next we place the remaining safety. For each scenario, the remaining safety may be placed at the deep level in exactly one way and this will respect the self symmetries. Next we must count the number of formations with the safety at the mid level.

1. The mid level safety can be placed outside all linebackers or in between two linebackers (ss) for a total of two ways.

2. For parts (a) and (b) which are not self-symmetric, the mid level safety can be placed on either side of the linebacker for a total of two ways. Because (c) is self-symmetric we apply mirror symmetry and place the safety in one way creating a new formation which is not self-symmetric.

3. For all formations the mid level safety is placed as the lone player at the mid level one way. This gives two new self-symmetric formations and four new alignments which are not self-symmetric.

We recall a defense will be (ss) if and only if all levels of the defense are (ss), meaning that we have four self-symmetric formations with both safeties deep and exactly $1 \cdot 1 + 1 \cdot 2 = 3$ self-symmetric formations with one safety at the mid level. For the non-self-symmetric formations we have six formations with both safeties deep and with a safety at the mid level we have $1 \cdot 1 + [2 \cdot 2 + 1 \cdot 1] + 4 \cdot 1 = 10$ formations. Thus we have seven self-symmetric formations and 16 non-self-symmetric formations.

Lastly we must place the remaining three cornerbacks. Since three cornerbacks will be on the field no alignment of them will be self-symmetric and from our restrictions a cornerback can line up at any level of the defense giving each one four possible levels and we come to three scenarios.

1. All three cornerbacks line up on the same side and same level of the field. As with the above calculations, there are eight total possible alignments and four with respect to the mirror equivalence.

2. Exactly two cornerbacks line up on the same side and level of the field. This gives eight possible regions for the pair of cornerbacks and seven for the lone cornerback yielding 56 total formations. If we take into account the mirror equivalence, each formation is equivalent to exactly one other so we have $56/2 = 28$ distinct alignments.

3. All three cornerbacks line up in distinct regions of the field. This can be done in $\binom{8}{3} = 56$ ways without regard to the mirror equivalence. Each formation is mirror equivalent to exactly one other so we have $56/2 = 28$ distinct alignments.

Thus there is a total of $8 + 56 + 56 = 120$ formations and with respect to the mirror equivalence there are $4 + 28 + 28 = 60$ distinct formations. Thus we have $7 \cdot 60 + 16 \cdot 120 = 2,340$ ways to align the nickel defense.

The Dime Defense

The final defensive formation under consideration, the dime, is used exclusively against offensive passing plays. The dime formation has exactly four corners and two safeties with at least one linebacker. See Figure 8.4 for a sample formation. By the restrictions, one of the safeties is fixed at the deep level and there are at least two linemen on the field. We have the following cases:

1. four linemen and one linebacker

2. three linemen and two linebackers

3. two linemen and three linebackers.

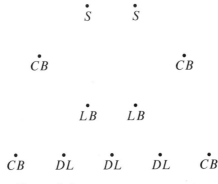

Figure 8.4. A standard Dime formation

We place the linebackers on the defensive line or the mid level.

1. (a) The linebacker stands on mid level (ss) one way.
   (b) The linebacker stands on defensive line level three ways; outside, inside with two linemen on either side (ss), and inside with one lineman on one side and three on the other side.

2. (a) Both linebackers are on the mid level (ss) one way.
   (b) One linebacker is on the mid level and the other is on the defensive line level in two ways, inside or outside on the defensive line(neither is ss).
   (c) Both linebackers are on the defensive line level three ways; both outside (ss), both inside (ss), and one inside and one outside in two ways, that is, with one lineman between the linebacker or two.

3. (a) Three linebackers on the mid level (ss) one way
   (b) Two linebackers on the mid level and one on the defensive line level in two ways, inside (ss) in one way and outside in one way.
   (c) One linebacker on mid level and two on defensive line level in two ways, both outside (ss) and one inside and one outside.

We place the second safety. Again he can stand on the defensive line level or the mid level. If both safeties are deep we have eight self-symmetric formations and eight formations which are not self-symmetric. We consider the case where the second safety is on the mid level.

1. (a) With symmetry the safety can stand on the mid level next to the linebacker one way.

(b) The safety stands alone on the mid level one way forming a new self-symmetric formation and two others which are not.

2. (a) The safety can be placed on mid level in two ways, between the two linebackers (ss) and outside the linebackers.

   (b) In either formation the safety can stand next to the linebackers in two ways, that is, to the left or right.

   (c) For all alignments the safety stands alone on the mid level in one way. We have two new self-symmetric formations and two new alignments which are not self-symmetric.

3. (a) The safety can stand in two ways, inside the linebackers and outside the linebackers.

   (b) For the self-symmetric alignment, the safety can stand in two ways, between the linebackers (ss) and outside the linebackers. The other formation is not self-symmetric, so the safety can stand in three ways, that is, between the linebackers or on either side. We have created one self-symmetric formation and four new formations which are not self-symmetric.

   (c) Considering the formation which is self-symmetric, the safety can stand one way outside the linebackers. The other formation is not self-symmetric, so the safety can stand in two ways outside the linebackers, giving three new formations which are not self-symmetric.

Recall there are eight alignments that are self-symmetric and eight alignments that are not when both safeties are deep. Including these formations, to place the seven players in the middle of the field, we have

$$8 + 1 \cdot 1 + 1 \cdot 2 + 1 \cdot 1 + 2 \cdot 2 + 1 \cdot 2 + 2 \cdot 1 + 1 \cdot 1 + 3 \cdot 1 + 1 \cdot 1 + 2 \cdot 1 = 27$$

formations that are not self-symmetric and

$$8 + 1 \cdot 1 + 1 \cdot 1 + 1 \cdot 2 + 1 \cdot 1 = 13$$

formations that are.

It is left to count the number of ways to place four cornerbacks. Since we allow a corner the freedom to line up on the same side and line of the field, we have five cases.

1. All four cornerbacks stand on the same level on the same side of the field. There are four levels and two sides, so this can be accomplished eight ways without mirror symmetry. Otherwise each formation is equivalent to exactly one other, so under mirror equivalence we thus have four formations.

2. Exactly three cornerbacks stand on the same level and on the same side of the field. Choose from eight regions to place the three standing together and seven to place the other cornerback. We have 56 formations without regard to mirror equivalence. Each formation is mirror symmetric to one other, so we have $8 \cdot 7/2 = 28$ formations.

3. We form two groups of exactly two cornerbacks standing on the same level and same side of the field. This is equivalent to counting the placement of two cornerbacks standing on either a different level or a different side of the field. There are $\binom{8}{2} = 28$ ways to place these two groups. Four of these formations are self-symmetric and the rest are equivalent to exactly one other. We have 4+12=16 formations with regard to mirror symmetry.

4. One group of two cornerbacks stands in the same region and the other two single cornerbacks stand in distinct regions. Place the set of two in eight ways. There are $\binom{7}{2} = 21$ ways to place the other two cornerbacks. Again each of these formations is mirror symmetric to exactly one other. We have $8 \cdot 21/2 = 84$ possible formations.

5. All four cornerbacks stand in distinct regions of the field. This can be done in $\binom{8}{4} = 70$ ways. There are $\binom{4}{2} = 6$ that are self-symmetric. The other 64 have a reflection and are symmetric to exactly one other. We have 6+32=38 formations after applying mirror equivalence.

Thus there are
$$4 + 28 + 16 + 84 + 38 = 170$$
ways to place the four cornerbacks with mirror symmetry and
$$8 + 56 + 28 + 168 + 70 = 330$$
ways to place the cornerbacks without regard to mirror symmetry, so we have
$$170 \cdot 13 + 330 \cdot 27 = 11,120$$
total ways to build a dime defense.

8.4 Conclusion

After the calculations, the Patriots' fan bursts out that there are exactly 15,372 possible defensive alignments for Belichick to choose from. During her exclamation of how amazing it is for one coach to be able to choose exactly the correct defense in 30 seconds, the last play of the game begins to unfold. The Pittsburgh quarterback hikes the ball, scans the opposing defense, recognizes the formation, and decides the best plan of action. He sets his feet and heaves the ball towards the endzone. The fans wonder if Belichick's decision was the right one for their team.

About the Authors

Tricia Muldoon Brown is an Assistant Professor of Mathematics at Armstrong Atlantic State University. Her research interests include algebraic and topological combinatorics, specifically studying the order complexes of certain poset products. After obtaining her doctorate at the University of Kentucky, and in the process becoming an avid fan of the Wildcats, Tricia and her husband moved south to new jobs in Savannah. When not teaching or researching, Tricia enjoys reading, walking her dogs, playing fantasy football, and cheering for the Steelers.

Eric Brendan Kahn is an Assistant Professor of Mathematics at Bloomsburg University. He wrote his dissertation on a topic from group theory that was motivated by algebraic topology while at the University of Kentucky. It was in Lexington that he developed his passion for teaching and decided to pursue his career at a teaching university. Aside from academia, Eric enjoys cooking, working out, reading, actively cheering for Boston's professional sports teams, and spending time with his wife and dog.

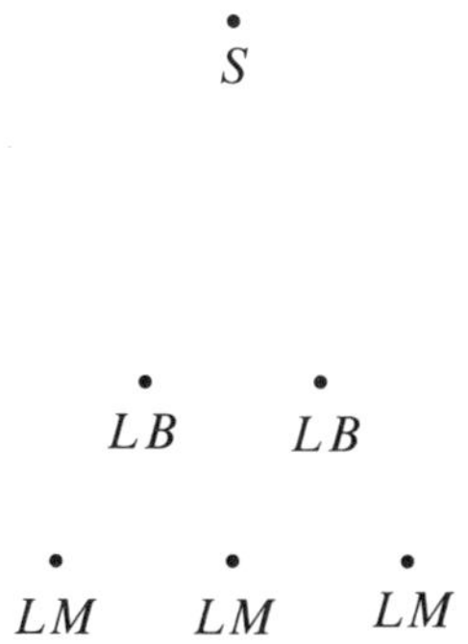

•
S

• •
L B   *L B*

• • •
LM   *LM*   *LM*

Figure 8.5. Fixed Personal in 3-4

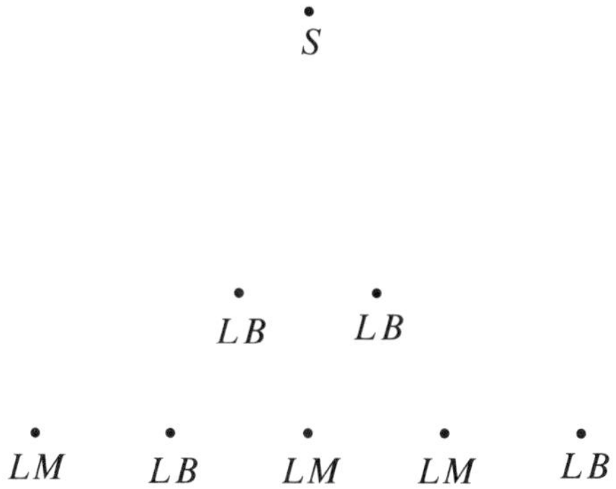

•
S

• •
L B   *L B*

• • • • •
LM   *L B*   *LM*   *LM*   *L B*

Figure 8.6. 3-4 Defense at Stage 1b. Alignment

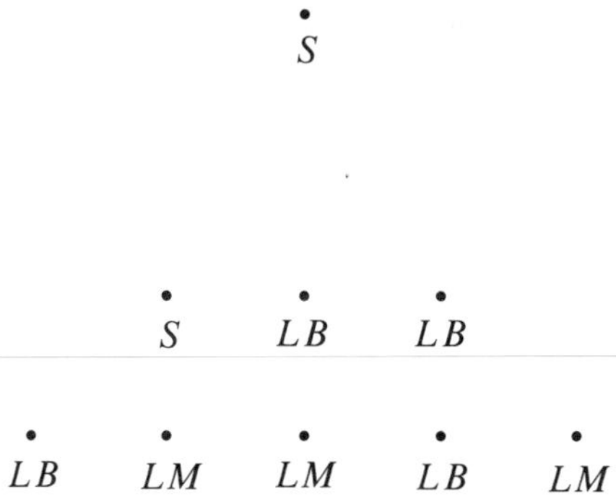

•
S

• • •
S   *L B*   *L B*

• • • • •
L B   *LM*   *LM*   *L B*   *LM*

Figure 8.7. Interior 3-4 Defense in 1b. Alignment with the Safety at the Mid Level

CHAPTER 9

A Look at Overtime in the NFL

Chris Jones

9.1   Introduction

There is debate as to how NFL overtime games should be decided. In the current system a coin toss takes place and the team that wins has the choice to either kick-off or receive the ball. They can also defer this choice and choose an end of the field to defend. The game is played with regular NFL rules and the first team to score wins. In a regular season game the teams play for fifteen minutes, and if neither team scores the game is declared a tie. In the playoffs, where there must be a winner, they keep playing until someone scores.

Ideally, the winner in overtime should be independent of which team wins the toss, but this is not the case. In the period 2004–2008, 72 regular-season games went into overtime. Of those, 46 were won by the team that received the overtime kick-off, giving a success rate of 64% for the team winning the toss (with one exception, teams always opt to receive rather than kick in overtime.) Over this period the loser of the coin toss won 25 games or just 35% of the time, with one game ending tied. Thus the coin flip is important in determining the winning team. Many games are won by the team receiving the ball before the opposition's offense even gets a chance to go onto the field. Of the 72 overtime games, 28, or 39% were won without the kicking team's offense touching the ball.

Using Markov chains we can verify that, given the likelihood of a score on any given possession, this success rate for the receiving team is what we might expect. We then look at an alternative method for deciding the winner of overtime games and how it could reduce (although not eliminate) the advantage afforded by the coin toss.

9.2 Game Data

In the 2008 season there were 5461 total possessions not ended by the completion of a half. Of them, 1122 ended in touchdowns by the offense and 845 ended in field goals. Thus $1967/5461 = 36\%$ of possessions resulted in the offense scoring. This fits well the numbers from 2004–08, where 39% of overtime games are won with only one team ever having the ball. The difference between the number of possessions ending in a score (36%) and those overtime games ending on one possession (39%) can be explained by the more conservative approach an offense can take close to the goal, knowing that a field goal is a possibility.

Of the 5461 possessions, 2294 ended in punts, in 231 the ball was turned over on downs, 155 field goal attempts were missed, and 708 resulted in non-scoring turnovers: fumbles and interceptions that were not returned for touchdowns. A total of 21, 52, and 33 possessions ended in safeties, interceptions returned for touchdowns, and fumbles returned for touchdowns respectively. Thus in any given possession we can assume the following probabilities:

36% result in the offense winning.

$$\frac{2294 + 231 + 708 + 155}{5461} = 0.62 \text{ or } 62\% \text{ result in a change of possession.}$$

$$\frac{21 + 52 + 33}{5461} = 0.019 \text{ or } 2\% \text{ result in the defense winning}$$
via a turnover or safety.

9.3 Analyzing the current system

Let us assume we have two teams, A and B and that team A receives the overtime kick-off. We can create a Markov chain matrix for our four possible states. These states listed in order are "Possession for Team A", "Possession for Team B", "Team A won" and "Team B won." Table 9.1 shows the probability of going from each row state to each column state. The first row of the matrix T_1 tells us the probability of being in any given situation after one possession, assuming team A starts with the ball.

$$T_1 = \begin{bmatrix} 0 & 0.62 & 0.36 & 0.02 \\ 0.62 & 0 & 0.02 & 0.36 \\ 0 & 0 & 1 & 0 \\ 0 & 0 & 0 & 1 \end{bmatrix}.$$

Table 9.1.

	Possession for A	Possession for B	A wins	B wins
Possession for A	0	0.62	0.36	0.02
Possession for B	0.62	0	0.02	0.36
A wins	0	0	1	0
B wins	0	0	0	1

To find the probabilities after n possessions we take the matrix above to the nth power and again read along the top row. For example, after three possessions,

$$T_1^3 = \begin{bmatrix} 0 & 0.24 & 0.51 & 0.25 \\ 0.24 & 0 & 0.25 & 0.51 \\ 0 & 0 & 1 & 0 \\ 0 & 0 & 0 & 1 \end{bmatrix}.$$

Which says that after three possessions the probability of team A having won is 0.51, while the probability of team B having won is 0.25. This leaves a 0.24 probability that the game is still going and as we have completed three alternating possessions, team B has the ball, hence the entry in the second column of the first row.

Now we can look at how a game should finish given our analysis. During the 2008 regular season, there were approximately six possessions per quarter, so we assume that the teams play until the end of six possessions. From

$$T_1^6 = \begin{bmatrix} 0.06 & 0 & 0.57 & 0.37 \\ 0 & 0.06 & 0.37 & 0.57 \\ 0 & 0 & 1 & 0 \\ 0 & 0 & 0 & 1 \end{bmatrix}$$

we see that there is a 6% chance of the game not having been completed, a 57% chance the receiving team wins, and a 37% chance the kicking team wins. This appears to overestimate the probability of a game not being completed because only once in 2004-2099 did a game end in a tie. This is no doubt caused by the effect previously mentioned whereby a team close to the goal can kick a field goal, safe in the knowledge that they only need three points to win and avoid the risk of a turnover close to the endzone while going for a touchdown. If we adjust the matrix to fit the data, we can reduce the probability of a possession change and increase the probability of points being scored on any given possession.

By trial and error we obtain the matrix,

$$T_2 = \begin{bmatrix} 0 & 0.5 & 0.48 & 0.02 \\ 0.5 & 0 & 0.02 & 0.48 \\ 0 & 0 & 1 & 0 \\ 0 & 0 & 0 & 1 \end{bmatrix}$$

which when taken to the power six gives

$$T_2^6 = \begin{bmatrix} 0.02 & 0 & 0.64 & 0.34 \\ 0 & 0.02 & 0.34 & 0.64 \\ 0 & 0 & 1 & 0 \\ 0 & 0 & 0 & 1 \end{bmatrix}.$$

This more accurately matches the results of overtime games over the five season period. The matrix states there is a 64% chance team A has won and a 34% chance B has won, with a 2% chance of a tie, which compares to 64%, 35%, and 1% from the actual games. However the estimate says that roughly 50% of games should be won without the kicking team getting the ball, which does not agree with the actual number of 39%.

9.4 An alternative proposal

We now consider a proposal that has been made to reduce the advantage of the coin toss. As we have seen, the sudden-death method gives a significant advantage to the team that receives the ball. One proposal has been that rather than playing until one team scores, the teams play with the winner being the first to score six points. A game can still be won on one possession, but now it requires a touchdown rather than a field goal. If no touchdown is scored the winning team can score two field goals to win the game. While this does seem likely to balance out the advantage somewhat, we will see that the team that receives the ball does still have an advantage and now the game will, in general, go longer, either resulting in potentially long playoff games, or more ties from games in which the teams haven't reached the requisite six points by the end of fifteen minutes.

Teams act differently when they only need a field goal to win. Thus we shall assign different probabilities for scoring that depend on a team's situation. If a team has no points we shall take the probabilities from the games. The probabilities of a touchdown, field goal, change of possession, and op-position touchdown are given in Table 9.2.

The change of possession includes punts, turnover on downs, missed field goals, non-scoring turnovers, and safeties. While a safety does not naturally

Table 9.2.

Outcome	Probability
Touchdown	$\dfrac{1122}{5461} = 0.21$
Field Goal	$\dfrac{845}{5461} = 0.15$
Turnover for touchdown	$\dfrac{52 + 33}{5461} = 0.02$
Change of possession	$\dfrac{2294 + 231 + 155 + 708 + 21}{5461} = 0.62$

fit into this category, a single safety essentially means only a change of possession as a team would still need two field goals or a touchdown to win, and the chance of two safeties for one team in a single overtime is negligible.

If a team already has three points, then it can switch to the conservative approach we believed they were using in the original overtime. As in matrix T_2 we now estimate that the probability of a score of any kind for the team in possession is 0.48, the probability of a change of possession is 0.5, leaving a probability of 0.02 of the defense scoring a touchdown. Summarizing the probabilities is a table, where the team in possession and the current score (with team A's points first) are listed, we get Table 9.3.

Table 9.3.

	A 0-0	B 0-0	A 3-0	B 3-0	A 0-3	B 0-3	A 3-3	B 3-3	A won	B won
A 0-0	0	0.62	0	0.15	0	0	0	0	0.21	0.02
B 0-0	0.62	0	0	0	0.15	0	0	0	0.02	0.21
A 3-0	0	0	0	0.5	0	0	0	0	0.48	0.02
B 3-0	0	0	0.62	0	0	0	0.15	0	0.02	0.21
A 0-3	0	0	0	0	0	0.62	0	0.15	0.21	0.02
B 0-3	0	0	0	0	0.5	0	0	0	0.02	0.48
A 3-3	0	0	0	0	0	0	0	0.5	0.48	0.02
B 3-3	0	0	0	0	0	0	0.5	0	0.02	0.48
A won	0	0	0	0	0	0	0	0	1	0
B won	0	0	0	0	0	0	0	0	0	1

Putting these values into a Markov chain matrix T_3 and raising it to the sixth power we have

$$
T_3^6 = \begin{bmatrix}
0.06 & 0 & 0.04 & 0 & 0.04 & 0 & 0.02 & 0 & 0.48 & 0.37 \\
0 & 0.06 & 0 & 0.4 & 0 & 0.04 & 0 & 0.02 & 0.37 & 0.48 \\
0 & 0 & 0.03 & 0 & 0 & 0 & 0.02 & 0 & 0.75 & 0.20 \\
0 & 0 & 0 & 0.03 & 0 & 0 & 0 & 0.02 & 0.58 & 0.37 \\
0 & 0 & 0 & 0 & 0.03 & 0 & 0.02 & 0 & 0.37 & 0.58 \\
0 & 0 & 0 & 0 & 0 & 0.03 & 0 & 0.02 & 0.20 & 0.75 \\
0 & 0 & 0 & 0 & 0 & 0 & 0.02 & 0 & 0.64 & 0.34 \\
0 & 0 & 0 & 0 & 0 & 0 & 0 & 0.02 & 0.34 & 0.64 \\
0 & 0 & 0 & 0 & 0 & 0 & 0 & 0 & 1 & 0 \\
0 & 0 & 0 & 0 & 0 & 0 & 0 & 0 & 0 & 1
\end{bmatrix}.
$$

Assuming that A started with the ball, the top row tells us the probability that A has won after six possessions is 48% compared to 57% in the previous model. B's chance's have increased from 34% to 37%. Unfortunately, and unsurprisingly, a great deal of the drop in A's chance of winning comes not from an increase in team B's chance of winning, but from a larger chance of a game not being completed, which is now 15%, compared to just 6% previously. Given our new rule, it is also more difficult for team B to win within six possessions, hence the increase in ties.

Finally we can look at how a playoff game, one with no time limit, may turn out in the new system. We have

$$
\lim_{n \to \infty} T_3^n = \begin{bmatrix}
0 & 0 & 0 & 0 & 0 & 0 & 0 & 0 & 0.57 & 0.43 \\
0 & 0 & 0 & 0 & 0 & 0 & 0 & 0 & 0.43 & 0.57 \\
0 & 0 & 0 & 0 & 0 & 0 & 0 & 0 & 0.78 & 0.22 \\
0 & 0 & 0 & 0 & 0 & 0 & 0 & 0 & 0.60 & 0.40 \\
0 & 0 & 0 & 0 & 0 & 0 & 0 & 0 & 0.40 & 0.60 \\
0 & 0 & 0 & 0 & 0 & 0 & 0 & 0 & 0.22 & 0.78 \\
0 & 0 & 0 & 0 & 0 & 0 & 0 & 0 & 0.65 & 0.35 \\
0 & 0 & 0 & 0 & 0 & 0 & 0 & 0 & 0.35 & 0.65 \\
0 & 0 & 0 & 0 & 0 & 0 & 0 & 0 & 1 & 0 \\
0 & 0 & 0 & 0 & 0 & 0 & 0 & 0 & 0 & 1
\end{bmatrix}.
$$

The advantage to the receiving team is 57-43. This compares well with the 65-35 advantage we came up with in the first case (the 65-35 advantage occurs in rows 7 and 8 when we begin at 3-3, which is identical to the position we began with previously when any score will win.)

Table 9.4.

Number of of possessions	Chance of completion
6	85%
8	93%
9	95%
10	97%
12	99%

A final consideration is the length of the game. While we have said that length is irrelevant in an overtime game, it is clearly of concern to the NFL for television scheduling.

The probabilities of the game being completed after a given number of possessions are in Table 9.4. It takes us around twelve possessions, or one half of a football game to reach a 99% chance of the game having finished.

9.5 Conclusion

The conclusion we reach is not surprising. By requiring a total score of six points, the average length of games will increase. Our T_2 matrix, in which there is a 0.5 probability of a team winning on any given possession, gives an expected number of possessions of an overtime game of $\sum_{n=1}^{\infty} n(0.5^n) = 2$. In the new system the expected number of possessions is a much higher 9.3. Given the undesirability of a regular season game going beyond one quarter in overtime, and the lack of appetite among sports fans for games finishing in a tie, it seems unlikely that this new system would be beneficial for regular season games. However, one can see an appeal for a postseason playoff game. Not only is the advantage of the toss reduced, which has to be good for fans who want the best team to win, but the fact that the game may go longer may not be such a bad thing for ratings-hungry television networks.

About the Author

Chris Jones received his BSc and PhD from the University of Salford, England having chosen the closest University to Manchester United for his study. His primary area of research is coding theory, which led him to a postdoc at the University of Virginia. He now teaches math at St. Mary's College of California in the Bay Area. He enjoys most sports and has discovered that a childhood love of cricket transfers nicely to baseball.

CHAPTER 10

Extending the Colley Method to Generate Predictive Football Rankings

R. Drew Pasteur

Abstract

Among the many mathematical ranking systems published in college football, the method used by Wes Colley is notable for its elegance. It involves setting up a matrix system in a relatively simple way, then solving it to determine a ranking. However, the Colley rankings are not particularly strong at predicting the outcomes of future games. We discuss the reasons why ranking college football teams is difficult, namely weak connections (as 120 teams each play 11–14 games) and divergent strengths-of-schedule. Then, we attempt to extend this method to improve the predictive quality, partially by applying margin-of-victory and home-field advantage in a logical manner. Each team's games are weighted unequally, to emphasize the outcome of the most informative games. This extension of the Colley method is developed in detail, and its predictive accuracy during a recent season is assessed.

Many algorithmic ranking systems in collegiate American football publish their results online each season. Kenneth Massey compares the results of over one hundred such systems (see [9]), and David Wilson's site [14] lists many rankings by category. A variety of methods are used, and some are dependent on complex tools from statistics or mathematics. For example, Massey's ratings [10] use maximum likelihood estimation. A few methods, including those of Richard Billingsley [3], are computed recursively, so that each week's ratings are a function of the previous week's ratings and new results. Some high-profile rankings, such as those of USA Today oddsmaker Jeff Sagarin [12], use methods that are not fully disclosed, for proprietary reasons. Despite the different approaches, nearly all ranking methods use the same simple data set, the scores of games played during the current season. A few also use other statistics, such as yardage gained. College football's Bowl Championship Series (BCS), which matches top teams in financially-lucrative postseason games, including an unofficial national championship game, chooses its teams using a hybrid ranking, currently including two human polls and six computer rankings [5]. The use of victory margins in BCS-affiliated computer rankings was prohibited following the 2001 season [4], out of concern that coaches might run up huge margins, violating good sportsmanship.

There are two opposing philosophies in ranking methods, leading to *retrodictive* and *predictive* rankings. Retrodictive rankings aim to reflect most accurately the results of the current season in hindsight (minimizing *violations*, cases of a lower-ranked team defeating a higher-ranked one). Predictive rankings attempt to identify the strongest teams at the present, so as to forecast the winners of upcoming contests. Most predictive rankings use margin-of-victory and also consider home-field advantage, both in previous and future games. To achieve reasonable early-season results, predictive ranking methods typically carry over data from the previous season, perhaps with adjustments made for returning or departing players and coaching changes. Retrodictive rankings start from scratch each season, so they are not published until each team has played several games. Jay Coleman's MinV ranking [6, 7], designed to achieve optimal retrodictive results, is superior to any other ranking in that category [9]. No predictive ranking algorithm has consistently outperformed the Las Vegas oddsmakers [1], who have a strong financial interest in accurately assessing upcoming games. Many ranking systems seek a balance of predictive and retrodictive quality, attempting to give insight into future contests while remaining faithful to past results.

Among the six Bowl Championship Series computer rankings, the Colley Matrix, designed by astrophysicist Wes Colley, is unique. Colley's algorithm [8], rooted in linear algebra, is elegant in its simplicity. His rankings are relatively easy to reproduce, and Colley's method involves neither margin-of-victory nor home-field advantage. This method is not a strong predictor [2], nor is it designed to be, raising the question of whether it could be extended to create a predictive algorithm based on linear algebra. In such an attempt, we must accept losing the elegance of Colley's original method, and will need to use additional data, such as victory margins and home-field advantage.

To extend Colley's method in this way, we must first understand how it works. The method begins by rating each team using a modified winning percentage, then adjusts the ratings according to the quality of opposition a team has faced. Each team effectively starts the season with a win and a loss, to avoid having undefeated teams automatically at the top (and winless teams at the bottom) regardless of schedule strength. All teams are placed in a fixed, arbitrary order, so that each row of a square matrix system $Ax = b$ relates to a particular team. Before any games are played, A is a diagonal matrix with 2's for all diagonal entries, and b is a vector consisting of all ones, as shown in (10.1). If we solve the system, we find that $x_j = 0.5$ for each j, agreeing with a winning percentage of 0.5, from an implied 1-1 initial record. Thus, all teams are considered equal prior to the start of the season.

$$
\begin{pmatrix}
2 & & & & \\
& 2 & & & \\
& & 2 & & \\
& & & \cdots & \\
& & & & 2
\end{pmatrix}
\begin{pmatrix}
x_1 \\ x_2 \\ x_3 \\ \vdots \\ x_n
\end{pmatrix}
=
\begin{pmatrix}
1 \\ 1 \\ 1 \\ \vdots \\ 1
\end{pmatrix}
\tag{10.1}
$$

To include the result of a game, we add one to the diagonal element associated with each of the teams involved, and subtract one from each of the two off-diagonal elements whose locations are coordinates are given (in either order) by the index numbers of those two teams. Finally, we add one-half to the entry of b associated with the winning team, and subtract one-half from the entry of the losing team. After all games have been included, the nonzero entries in A are as follows:

$a_{jj} = 2 +$ (number of games played by team #j)

$a_{ij} = -1 \cdot$ (number of games between team #i and team #j), for $i \neq j$

$b_j = \dfrac{1}{2} \cdot \left[2 + (\text{wins by team \#}j) - (\text{losses by team \#}j) \right]$

For example, if team #3 defeats team #1 in the season's first game, we would obtain the system in (10.2). The principle is that the rating of team #3 minus that of team #1 should equal one-half, assuming equal schedules, and our changes to each team's row reflect such a condition. Solving this system would show that $x_1 = 0.375$ and $x_3 = 0.625$ (with all other x_j still 0.5), so team #3 is currently rated as the strongest team, while team #1 is rated the weakest. This is logical, given that we have no information regarding the strength of any other teams. The strength-of-schedule inequality reduces the difference between the teams' ratings to $0.625 - 0.375 = 0.25$, instead of the original 0.5 difference.

$$
\begin{pmatrix}
3 & -1 & & & \\
 & 2 & & & \\
-1 & 3 & & & \\
 & & & \ddots & \\
 & & & & 2
\end{pmatrix}
\begin{pmatrix}
x_1 \\ x_2 \\ x_3 \\ \vdots \\ x_n
\end{pmatrix}
\begin{pmatrix}
0.5 \\ 1 \\ 1.5 \\ \vdots \\ 1
\end{pmatrix}. \tag{10.2}
$$

At any time, we can solve $Ax = b$ to determine the team rating vector x, and rank the teams according to their x_j values, in descending order. For future games, we would predict a team with a higher rating (from x) to defeat any with a lower rating.

One of the difficulties in algorithmic college football rankings is that there are many teams (120 in the highest classification, and a total of over 700 NCAA teams), yet each team plays relatively few games in a season, typically 10–13. Of these games, 7–8 games are played within a conference, and non-conference opponents are often chosen partially by geographic proximity. Most randomly chosen pairs of teams do not play one another, but are connected only through chains of common opponents. By season's end, any two of the 120 teams can be connected by three or fewer intermediate teams. (In graph theory terms, this is equivalent to a graph diameter of four.) However, the early-season lack of connectedness increases the difficulty of ranking teams. College basketball, despite having more teams, presents a simpler ranking problem, because there are more non-conference games and more interregional contests. By their design, some methods handle such weakly-connected networks better than others, and Colley notes that the performance of his method is dependent on the degree of connectedness among the teams [8]. While this cannot be completely resolved without fundamentally changing the model's structure, we will attempt to compensate for it.

Another ranking issue is strength-of-schedule differential, which largely arises from conference affiliations, since most teams play roughly two-thirds of their games against league foes. Of the eleven conferences in major college football, three (the Sun Belt, Mid-American, and Conference USA) are relatively weak. Over the last five seasons, each of these leagues has won fewer than one in seven games against teams from the BCS-affiliated conferences [13]. It is common for a team in such a conference to go through an entire season without playing any nationally-ranked opponents, so even an undefeated record might not be meaningful. On the other hand, teams in the Big 12 and Southeastern Conference routinely play as many as six or seven regular-season games against consensus top-25 teams. The issue of schedule differences is not difficult to address within the Colley framework.

We hope to extend the Colley model in a way that improves predictive accuracy. We will leave much of the structure intact, but will alter the coefficient changes made with when including each new game result. The changes are as follows:

- Include margin-of-victory

- Weight games unequally, depending both on margin-of-victory and the expected result.

- Weight recent games more heavily.

- Start each season with teams having unequal ratings, based on results from the previous season, to improve early-season predictions, but diminish the effects of the initial inequalities as more games are played.

- Quantify home-field advantage, for use in ratings-based predictions.

While margin-of-victory is no longer used in the BCS rankings, it remains useful as a predictive tool. A team that consistently wins by substantial margins is likely to be superior to one that wins by just a few points, unless there is a substantial difference in strength-of-schedule. A similar argument can be made regarding losing teams that are soundly defeated compared with those that often lose close games. Victory margins are important, but we must choose carefully how best to use them. Scoring a late touchdown to increase a small lead is significant, but once a team is far ahead, an additional score becomes largely irrelevant. Thus, we need a diminishing returns principle applied to victory margins, obtained by using a sigmoidal (S-shaped) curve to determine the margin-based output value for a particular game. We will use a cumulative normal distribution function, translated to pass through the origin, as shown in Figure 10.1.

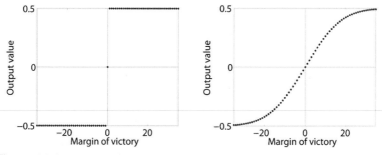

Figure 10.1. Output value m (to be added to b_j), based on the outcome of a game, under (left) the Colley system, and (right) a diminishing-returns principle.

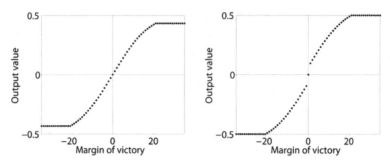

Figure 10.2. Output value m using a 21-point cap on margin-of-victory, before (left) and after (right) making a range-adjusting translation.

Some ranking systems that use margin-of-victory impose a cap, often 21 or 28 points, the equivalent of 3 or 4 touchdowns, ignoring any margin in excess of that cap, and we will use a 21-point cap. This results in a function with a range smaller than the desired $[-0.5, 0.5]$ range. One way to resolve this issue would be to multiply the capped function by a constant to stretch it vertically. We will take a different approach, inserting the missing increment between the zero point (representing a tie game) and ± 1 point values, as seen in Figure 10.2. The translations offer a middle ground in between a margin-blind formula and a fully margin-based formula.

In college football, games in which one team is very heavily favored (by 35 or more points) are not unusual, due to the large disparity between the best and worst teams. Using the Las Vegas odds, there were fourteen such games in 2008 [11]. These games may produce a no-win situation for the favored team under Colley's system, leading to a rating decrease regardless of the outcome, because the strength-of-schedule decrease can outweigh any benefit gained from adding a victory. On the other side, a low-rated team may be rewarded (with a rating increase) for playing a top opponent, regardless of how badly they are beaten. While this may be reasonable within some types of rating systems, (as they may aim to show which teams have performed the best against good competition) it is a weakness for predictive ratings. Logically, if a heavily-favored strong team wins easily over a weak opponent, then neither team's rating should change significantly as a result, because no new information was gained. Thus, we use differential weighting of games, based on both expected and actual outcomes. After we compute the new ratings, the expected outcome of games may change, so we re-weight every game and re-compute the ratings. This leads to a shift from direct solution (solving a matrix system $Ax = b$ once in Colley's method) to iterative computation, repeating steps until we obtain convergence, when teams' ratings stabilize. To compute the weight given to each game, we will apply the following principles:

- Heavily favored teams that win by substantial margins should not be penalized, so such games will have very small weights.

- Any true upset (a game in which the losing team is favored in hindsight) will receive the highest possible weight.

- A heavily-favored team that wins a close game will be penalized, but not as much as if they had lost the game.

- Barring upsets, games between evenly-matched teams are the most informative, so they will be weighted more heavily.

- The function computing game weights from margins should be smooth, while allowing a discontinuity at the point representing a margin of zero.

To follow these guidelines, we use a weight function that decreases exponentially with the product of the expected and actual margins, as shown in Figure 10.3, whenever the favored team wins. Both the favored team and the expected margin will be determined in hindsight, because of iterative computation.

For a variety of reasons, a team's performance may improve or decline during a season, i.e., we will weight recent games more heavily. To determine the weight w of a game played n weeks ago, we use the exponential

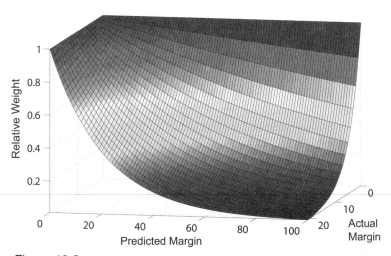

Figure 10.3. Relative weight function r for games won by the favored team.

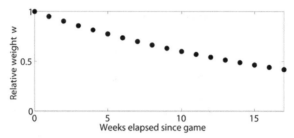

Figure 10.4. Time-based weight function w, using a 5% weekly decrease.

decay formula $w = \alpha^n$, where $0 < \alpha < 1$. Using $\alpha = 0.95$, we obtain time-based weights shown in Figure 10.4, which are multiplied by the margin-based weights to compute the weight for a particular game.

Just as every team starts a new season with a record of 0-0, regardless of the previous season's successes or failures, rankings should start fresh each year. In small, well-connected groups such as the National Football League, this is feasible, but in large, poorly-connected pools of teams, it is unrealistic. There may be few interregional games, and those may occur only in the playoffs, too late to be of predictive value. For this reason, many predictive systems use the results of one or more previous years as a starting point for a new season's rankings. As more data become available, the old results are weighted less, and perhaps dropped altogether once all teams are minimally connected. Using previous rankings as a starting point can improve predictive accuracy early in the season. If the starting values are not phased out entirely, they may help in predicting interregional games, particularly when certain regions have been historically strong or weak but scheduling is mostly localized. We will use the final ratings of the previous season as a starting point for a new year, and will consider them equivalent to two fully-weighted games, just as Colley in essence starts each team with a 1-1 record. Just as current-season games are weighted 5% less each week, we apply the same damping to the initial rating weights.

Computation of the value of home-field advantage could easily be the subject of a paper by itself. Some rating systems use a fixed value for the number of points by which a home team is improved, compared to playing on a neutral field. Others (such as Sagarin's ratings [12]) re-compute home-field advantage weekly during the season, either as a universal constant or with a value for each team. We will compute a single value based on previous seasons and use it throughout a given year. While college football teams rarely face an opponent twice during the same season, they routinely play the same team in consecutive years, with each team hosting one game. We will consider the margins of games in such home-and-home series to estimate a home-field advantage constant. We sum the margins, from the home team's perspective, in all such pairs of games over two seasons, then divide by the total number of games considered. By considering only home-and-home series, we eliminate home and away scheduling inequalities, such as the practice of top college teams paying weaker opponents to play at the stronger team's stadium with no return game. The home-field advantage value h was approximately 3.70 points in major-college football over two recent seasons. Since h should not change substantially in a single year, we will consider it constant through a season.

Putting together these ideas, we must compute two values used to change the matrix system when a game is added. First, the final weight y given to the game is a product of the relative weight r (based on expected and actual outcome) and time-based weight w. The coefficient adjustment z (applied to b_j) is a product of y and the margin-based game output value m.

$$y = rw, \quad z = ym \qquad (10.3)$$

We can then make the changes shown in (10.4) to account for a game in which team #3 defeats team #1. In (10.4), entries denoted by $*$ are unchanged.

$$\begin{pmatrix} +y & * & -y & & * \\ * & * & * & \cdots & * \\ -y & * & +y & & * \\ & & \vdots & \ddots & \\ * & * & * & & * \end{pmatrix} \begin{pmatrix} x_1 \\ x_2 \\ x_3 \\ \vdots \\ x_n \end{pmatrix} \begin{pmatrix} -z \\ * \\ +z \\ \vdots \\ * \end{pmatrix}. \qquad (10.4)$$

After solving the matrix system to obtain the rating vector $\mathbf{x}$, we can determine hypothetical margins of future (and past) games. To do this, we convert the values from x_j into ratings easily used for predictions. The conversion formula is given in (10.5), where c is a scaling factor (applied so that a one-point difference in ratings will be equivalent to a one-point predicted margin) and a is the desired rating for an average team. Based on data fitting experiments, we will use $c = 60$. While the value of a does not affect the predictive outcomes, we will use $a = 100$, keeping with common practice. Note that in (10.5), we subtract 0.5, the rating of an average team in the Colley system.

$$R_j = a + c \cdot (x_j - 0.5) \qquad (10.5)$$

To predict the outcome of a game, we consider the teams' ratings, R_h and R_v (for the home and visiting teams) given by (10.5). Unless the game is

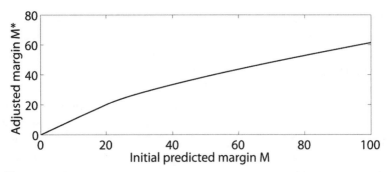

Figure 10.5. Adjustment of margin predictions which initially exceed 21 points.

contested at a neutral site, we add the home-field constant $h = 3.70$ points to the home team's rating. After this, the team with the higher rating will be predicted to win, and the difference between the modified ratings is the predicted margin-of-victory M, as seen in (10.6).

$$M = |R_h + h - R_v| \qquad (10.6)$$

For highly mismatched teams, the formula will typically overestimate the margin. A team that is 70 points better than its opponent is unlikely to win by such a huge margin, instead inserting reserve players and playing conservatively with a large lead. If the predicted margin M is greater than 21 points, we reduce it using the fractional-exponent formula given by (10.7). Based on data fitting, we take $\alpha = 0.8$.

$$M^* = 21 + \frac{1}{\alpha} \cdot \left[(M - 20)^{\alpha} - 1\right], \quad \text{if } M > 21. \qquad (10.7)$$

The margin, possibly adjusted, is then rounded to the nearest integer. Because tie games are not possible in college football, a margin of less than one-half point will be rounded up to one point instead of down to zero. In contrast, some rating systems (such as those of the Las Vegas oddsmakers) allow a predicted margin of zero.

The weight assigned to a game depends partly on its predicted outcome, so we need an initial set of ratings as a starting point. While we could initially rate all teams equally, we will instead begin with ratings carried over from the previous season. As we obtain new ratings, the predicted outcomes of some games will change (and for others the margin is altered without changing the predicted winner). These changes affect the weights assigned to the games, changing the ratings again. So, rating computation is an iterative process, in which ratings from one iteration are used to compute the game weights for the next. We continue until we obtain convergence, when ratings remain unchanged with additional iterations. Convergence usually takes only a few iterations.

The primary measure of a predictive rating system's quality is the percentage of games it correctly determines in advance; accuracy in determining margins is a secondary criterion. The Prediction Tracker [2] compares the results of dozens of rating systems. For the 2008 season, the system we describe correctly predicted the winner of 74.3% of games between two major-college (NCAA Division I FBS) teams. This compares favorably with other systems, placing it in the top 20% of all models. Prediction percentages for the second half of the season are also a good measure, because many ranking systems start from scratch each year, requiring more data to make accurate assessments. While more information is available later in the season, predictions become more difficult because conference games and postseason bowls are more often evenly-matched contests. Not surprisingly, our prediction percentage was lower during the latter half of the season, at 72.4%, but this still placed in the top quarter of all ranking systems. The final top 25 teams for the 2008 season, using our method, are listed as the Appendix.

There are many algorithmic rankings in major college football, probably because it has neither a playoff system nor an official national champion. However, I also have a strong interest in ranking high school football teams, dating back to several years spent teaching mathematics in a public high school. Power ratings and predictions for high school football in North Carolina and Ohio, computed by the same method discussed here, are published weekly during the season at www.fantastic50.net.

Some of the difficulties encountered in ranking college football teams are more apparent in those leagues. Ohio's 700+ teams each play only ten regular-season games, and North Carolina's geography leads to heavily-localized scheduling. These factors substantially reduce connectedness among the teams, so that at season's end, some pairs of teams are connected only through lengthy chains of intermediate teams. Strength-of-schedule differences can be enormous, as some large urban schools fielding three large teams play only against local peers, while small-school conferences may consist entirely of teams that have barely enough players to field one team. Obtaining scores of games involving rural schools with no nearby daily newspapers can be a challenge. However, high school football offers another interesting proving ground for mathematical ranking systems. Many states use a computational system of some kind to select and/or seed playoff teams, but there are few predictive rankings published in a given state. Such rankings offer a measuring stick for championship teams in lower classifications if they are unable to schedule games against the large-school powers.

Acknowledgements

Historical game scores and Las Vegas odds were obtained from Warren Repole; scores were also provided by Jay Coleman. Thanks to Eric Frantz of JJHuddle.com, Arnold Solomon of NCpreps.com, and Tim Stevens at The News and Observer, for kindly providing venues to publish the author's high school football rankings.

References

[1] T. Beck, College Football Ratings PT Awards, *The Prediction Tracker*, accessed on 16 July 2009. tbeck.freeshell.org/fb/awards2008.html

[2] ——, Computer Rating System Prediction Results for College Football, *The Prediction Tracker*, accessed on 16 July 2009.
www.thepredictiontracker.com/ncaaresults.php?type=1&year=08

[3] R. Billingsley, Dynamics of the system, Oct. 2008, *College Football Research Center*, accessed on 28 July 2009.
www.cfrc.com/Archives/Dynamics_08.htm

[4] ——, BCS Chronology, 25 June 2009, *Bowl Championship Series*, accessed on 11 August 2009. www.bcsfootball.org/bcsfb/history

[5] ——, BCS Computer Rankings, 11 Nov. 2008, *Bowl Championship Series*, accessed on 21 July 2009. www.bcsfootball.org/bcsfb/rankings

[6] J. Coleman, Minimum Violations (MinV) College Football Ranking, 10 Jan. 2009, *University of North Florida*, accessed on July 8, 2009.
www.unf.edu/~jcoleman/minv.htm

[7] ——, Minimizing Game Score Violations in College Football Rankings, *Interfaces* 35 (2005) 483–496.

[8] W. N. Colley, Colley's Bias Free College Football Ranking Method, 30 May 2002, *Colley Matrix*, accessed on 25 Aug. 2009.
www.colleyrankings.com/method.html

[9] K. Massey, College Football Ranking Comparison, 23 Mar. 2009, *Massey Ratings*, accessed on 22 July 2009. www.mratings.com/cf/compare.htm

[10] ——, Massey Ratings Description, 15 Aug. 2000, *Massey Ratings*, accessed on 28 July 2009. www.mratings.com/theory/massey.htm

[11] W. Repole, Historical Scores, 31 May 2009, *The Sunshine Forecast*, accessed on 23 June 2009. www.repole.com/sun4cast/scores2009.html

[12] J. Sagarin, Jeff Sagarin NCAA Football Ratings, 9 Jan. 2009, *USA Today*, accessed on 22 July 2009.
www.usatoday.com/sports/sagarin/fbt08.htm

[13] C. Stassen, Aggregate Conference Record, *Stassen.com College Football Information*, accessed on 20 Aug. 2009.
football.stassen.com/records/multi-conference.html

[14] D. L. Wilson, American College Football — Rankings, *Nutshell Sports Ratings*, accessed on 22 July 2009.
www.nutshellsports.com/wilson/popup.html

Appendix
Top 25 ranking, at the end of the 2008 season, by this method

Records shown include only games against FBS (formerly Division I-A) opponents.

1)	Florida (12-1)	14)	Virginia Tech (9-4)
2)	Southern Cal (12-1)	15)	Mississippi (8-4)
3)	Oklahoma (11-2)	16)	Oregon State (9-4)
4)	Texas (12-1)	17)	Florida State (7-4)
5)	Penn State (10-2)	18)	Missouri (9-4)
6)	Ohio State (9-3)	19)	California (9-4)
7)	Utah (12-0)	20)	Arizona (8-5)
8)	Oregon (10-3)	21)	West Virginia (8-4)
9)	Alabama (12-2)	22)	Iowa (8-4)
10)	Georgia (9-3)	23)	Oklahoma State (8-4)
11)	Boise State (11-1)	24)	Boston College (8-5)
12)	TCU (10-2)	25)	LSU (7-5)
13)	Texas Tech (9-2)		

About the Author

R. Drew Pasteur is an Assistant Professor of Mathematics at The College of Wooster. Before pursuing graduate study at North Carolina State University, he was a math teacher and head athletic trainer at Fuquay-Varina High School, in Fuquay-Varina, NC. It was during this time that Drew became interested in mathematical ranking algorithms and developed the first version of his Fantastic 50 high school football rankings. His other research area is in mathematical biology, applying dynamical systems to human physiology. Outside of mathematics, Drew enjoys singing, running, and spending time with his wife and son.

CHAPTER 11

When Perfect Isn't Good Enough: Retrodictive Rankings in College Football

R. Drew Pasteur

Abstract

Mathematical ranking systems, such as those used in college football's Bowl Championship Series (BCS), can be classified in two broad categories. Predictive methods seek to forecast outcomes of future games, while retrodictive rankings aim to most closely match the results of contests already played. Ideally, a retrodictive method would order teams such that each team is ranked ahead of all teams it defeated, and behind all the teams to which it lost. However, this is generally impossible at the end of the season, as any ranking will "violate" the results of some games, by having the loser ranked above the winner. For any given set of game results, there is a minimum possible number of violations, and we call a ranking that induces the minimal number a perfect ranking; computing such rankings is an NP-complete problem. Jay Coleman, an operations research professor at the University of North Florida, developed a model called MinV to identify perfect rankings. For the 2008 season, each of the six computer ranking systems used in the BCS had at least 80% more violations than MinV. However, perfect rankings are not unique, raising the question of which perfect ranking is the best ranking. If all perfect rankings agreed on the top teams, this might not be a major concern, but in the 2008 season, there were five teams that could have been #1 in a perfect postseason ranking. Because of clustered scheduling, it is possible to move groups of teams up or down, and sometimes even whole conferences, while maintaining a perfect ranking. Under MinV, a highly-ranked team may be unaffected by a loss to a far-inferior opponent, contrary to logic. The latter portion of this paper details multiple examples of these issues.

The highest division of collegiate football, the Football Bowl Subdivision (FBS) of the National Collegiate Athletic Association, formerly known as

Division I-A, is the only NCAA team sport that does not determine an official national champion. Based on long-standing tradition, top teams compete in season-ending bowl games, and a national champion is unofficially chosen by polls of coaches and media members. Several major conferences have historical ties to particular bowl games. For example, the champions of the Big Ten and Pac-10 conferences have played in the Rose Bowl nearly every year since 1947. More often than not, the consensus top two teams have not faced one another in a bowl game, sometimes leading to disagreements over which team was most deserving of the national title. Under this system, it was not uncommon for two major polls to name different national champions at season's end.

Since the formation of the Bowl Championship Series (BCS) before the 1999 season, the top two teams are guaranteed to face another in a *de facto* national championship game. The difficulty comes in determining which two teams should play in that game. Currently, the participants are selected by a formula that weights equally the results of the Harris Interactive College Football Poll [14], the USA Today Coaches' Poll [21], and a consensus of six independent computational rankings [1, 5, 10, 16, 19, 23]. Before 2004, other factors were included in the computation — losses, strength-of-schedule, and "quality" wins [6]. Because these items are implicitly included in the component rankings, their additional weighting was deemed unnecessary.

The six BCS computer rankings use a variety of algorithms to rank teams, but have in common that they use only the game scores and schedules to determine their rankings. Beginning with the 2002 season, computer ranking systems may not use margin of victory as a component factor [6]. While only these six computer rankings are used for BCS selections, many others exist. The Prediction Tracker [4] analyzes the accuracy of dozens of computer ranking systems, and David Wilson's website [22] lists and categorizes many more.

The methods used by the computational rankings vary widely. Richard Billingsley's rankings [5] are recursive, meaning that each week's rankings are computed using only the current week's games and the previous week's rankings. Some include margin-of-victory and/or home-field advantage, while others do not. While many involve advanced mathematical techniques, including maximum likelihood estimation (in Kenneth Massey's rankings [17]) or the solution of large linear systems (in the Colley Matrix rankings [11]), some use simple formulas. An example of a simple system is the Ratings Percentage Index (RPI), which ranks teams using a weighted average of winning percentage, opponents' winning percentage, and opponents' opponents winning percentage. While RPI is more commonly associated with college basketball, it is also widely published for college football.

There is no single measure of which ranking system is the best. Predictive ranking systems attempt to accurately forecast the outcome of upcoming games based upon currently available information. Teams are rated and ranked according to their perceived ability to win future games. Over the last decade, no computer rating system has consistently outperformed the Las Vegas oddsmakers. (See "most accurate predictor" in [3]). The bookmakers have a financial interest in accurate predictions, and ostensibly use all available information in setting the betting line, including injuries, detailed game statistics, and any relevant intangibles. At the opposite end of the spectrum from the predictive systems are retrodictive ranking systems, whose goal is to determine a ranking that accurately reflects the results of prior games. Such rankings do not use results from prior seasons, and most disregard margin-of-victory, so they are widely considered "fairer" for use in determining participants for postseason play. There is a middle ground between predictive and retrodictive rankings, as some rankings attempt to both model past results and predict future games.

If margins-of-victory and home-field advantage are removed, then evaluating the results of retrodictive ranking methods becomes simple. The best method is the one whose rankings contradict the game results least often. An ideal retrodictive method would determine a ranking with the property that each team is ranked above the teams they defeated and below the teams to which they lost. We will see that this is generally impossible, so the goal becomes to rank teams in a way that minimizes inconsistencies (known as *ranking violations* or *reversals*), occurrences in hindsight of a lower-ranked team beating a higher-ranked one. This is a variation of the *linear ordering problem* from graph theory, which involves ordering a group of objects such that some function is minimized, in this case the number of violations. The linear ordering problem is an example of an NP-complete problem ("Directed Optimal Linear Arrangement," denoted [GT43] in [12]), part of a class of computationally difficult problems for which no fast (polynomial-time), efficient, general method is known.

When the final 2008 rankings for the six BCS computer rankings are evaluated for retrodictive quality, the violation percentages are similar, as shown in Table 11.1. We will find that there is substantial room for retrodictive improvement in all of them. While zero violations would be ideal, there are multiple reasons why violations are inevitable.

In a three-team round-robin (where every team plays each of the others), it

Table 11.1. Ranking violation comparison among the six BCS computer rating systems.

BCS Rating System	Violations	Percentage
Sagarin	125	17.4%
Billingsley	126	17.6%
Massey	131	18.3%
Anderson/Hester	133	18.5%
Wolfe	133	18.5%
Colley	138	19.2%

is not uncommon for each team to win one game and lose one game, a situation called a *cyclic triad* [9] or a *loop* [20]. Such occurrences are noteworthy when the teams involved share a conference or divisional championship as a result, as was the case in the Big 12 South Division in 2008. Oklahoma beat Texas Tech, Texas Tech beat Texas, and Texas beat Oklahoma. Each team won all their other conference games, so these three teams had identical conference records (7-1), divisional records (4-1), and head-to-head records (1-1). The determination of which team advanced to the conference championship game had implications for the national championship. In a three-team loop, any ordering of the three teams will induce at least one violation, and some will cause two. While any of the three teams could be ranked highest relative to the others, the choice of the first team determines the order of the other two. To avoid a second ranking violation, the middle team must be the one which lost to the top team, so the lowest-ranked team is the one that defeated the top team. (Thus, the win of the lowest-ranked team over the highest-ranked team is the violation.)

There is another situation in which ranking violations are inevitable, the true upset, in which an inferior team defeats a superior one. A recent example of an upset is Mississippi's 2008 win over eventual national champion Florida. Mississippi had a record of 8-4, with losses to Wake Forest (8-5), Vanderbilt (7-6), South Carolina (6-6), and Alabama (12-2), while Florida (12-1) defeated three highly-rated teams and lost only to Mississippi. It is not reasonable to rank Mississippi ahead of Florida. Doing so would cause multiple violations, as Florida beat three of the teams that defeated Mississippi. (See Figure 11.1.) To avoid this, we accept the one violation induced by Mississippi's upset of Florida, and henceforth largely overlook this game.

Some unexpected results, especially those occurring early in the season, are not viewed as upsets in hindsight. Unheralded Alabama's season-opening

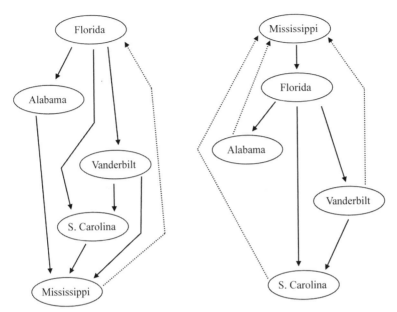

Figure 11.1. Two sets of relative rankings among teams, differing in how they handle Mississippi's victory over Florida. Teams nearest the top are ranked highest, and arrows indicate a victory by one team over another. Solid arrows indicate games in which the winner is ranked higher, while dashed arrows denote violations. In the ranking on the left, there is only one violation. In the ranking on the right, placing Mississippi above Florida leads to three violations.

victory against a preseason-top-ten Clemson team became less surprising with each passing week, as Alabama won its first twelve games while Clemson finished 7-6. Although Clemson was the higher-ranked team at the time of the game, making the result an upset in the traditional sense, we do not consider it an upset, because Alabama was ranked above Clemson by season's end. Upsets can be viewed as loops involving three or more teams. If Team A beats Team B, which beats Team C, which beats another, etc., and the last team beats Team A, then a loop is created.

A small number of necessary violations occur due to a pair of teams meeting twice during a season, with each team winning once. While major college football teams do not schedule two games against the same opponent during a given season, post-season contests (conference championships or bowl games) can produce rematches. One notable example was the Sugar Bowl following the 1996 season, in which Florida beat rival Florida State to win the national championship, after losing to Florida State in the last game of the regular season. When a pair of teams split two games, there will be exactly one violation, regardless of how the teams are ranked relative to one another.

Jay Coleman, an operations research professor at the University of North Florida, developed an integer programming model in [9] called *MinV* that determines the minimum possible number of ranking violations for a given set of data, then attempts to find a ranking which satisfies this minimum. The method used is beyond the scope of this paper, but the algorithm is described in [9] . We will consider a ranking to be retrodictively *perfect* if it results in the minimum number of violations. The problem of finding a perfect ranking is computationally intensive, but it is made slightly easier because perfect rankings are not unique. In practice, there are typically large numbers of perfect rankings, but they are still difficult to find among the $120! \approx 10^{200}$ permutations of the 120 teams in major-college football.

For each season since 2004, Coleman has published a post-season ranking [8] that has the minimum possible number of violations. For the 2008 season, there were 717 games contested between FBS teams, and the minimum number of violations was 69, so a perfect ranking correctly reflects the outcome of 90.4% of the games played during the season. By this measure of the retrodictive quality of a ranking system, Coleman's MinV rankings are far superior to other ranking systems, such as the BCS rankings in Table 11.1. Even the best of those six by this measure, the Sagarin ratings, has 81% more violations than Coleman's ranking. Of the 100+ rankings compared by Massey [15], only three others come within 25% of the minimum number of violations. Two of those three ranking systems are direct extensions of Coleman's MinV model; the other, GridMap [20], is independent, but uses similar principles.

Coleman's model is clearly outstanding at reflecting the outcome of games during the current season, so this raises the question of why nothing comparable to MinV is used in the BCS rankings. The answer lies in the non-uniqueness of perfect rankings. Because there are many possible perfect rankings, we would need further criteria to determine which one should be used. If the differences among top teams in perfect rankings are minor, then we might conclude that any such ranking is acceptable and could set arbitrary criteria to achieve uniqueness.

To determine the degree of variation among retrodictively perfect rankings, we consider the results of the 2008 season, including bowl games. Because only teams within the FBS are being ranked, games against non-FBS teams are not included in teams' records. To create new perfect rankings, we start with Coleman's final 2008 MinV ranking [8], then move teams so that

Table 11.2. Top-10 listings headed by each of the five possible #1 teams. In every case, teams #11–120 could remain unchanged from Coleman's MinV ratings.

1) Utah (12-0)	1) Florida (12-1)	1) Southern Cal (12-1)
2) Florida (12-1)	2) Southern Cal (12-1)	2) Florida (12-1)
3) Southern Cal (12-1)	3) Texas (12-1)	3) Texas (12-1)
4) Texas (12-1)	4) Oklahoma (11-2)	4) Oklahoma (11-2)
5) Oklahoma (11-2)	5) Utah (12-0)	5) Utah (12-0)
6) Alabama (12-2)	6) Alabama (12-2)	6) Alabama (12-2)
7) TCU (10-2)	7) TCU (10-2)	7) TCU (10-2)
8) Penn State (10-2)	8) Penn State (10-2)	8) Penn State (10-2)
9) Texas Tech (9-2)	9) Texas Tech (9-2)	9) Texas Tech (9-2)
10) Boise State (11-1)	10) Boise State (11-1)	10) Boise State (11-1)

1) Texas (12-1)	1) Texas Tech (9-2)
2) Florida (12-1)	2) Florida (12-1)
3) Southern Cal (12-1)	3) Southern Cal (12-1)
4) Oklahoma (11-2)	4) Texas (12-1)
5) Utah (12-0)	5) Oklahoma (11-2)
6) Alabama (12-2)	6) Utah (12-0)
7) TCU (10-2)	7) Alabama (12-2)
8) Penn State (10-2)	8) TCU (10-2)
9) Texas Tech (9-2)	9) Penn State (10-2)
10) Boise State (11-1)	10) Boise State (11-1)

no additional net violations are generated. Departing from Coleman's practice in the MinV rankings, we will not allow teams to be tied for a particular place.

There are five teams that could be #1 in a perfect ranking: Utah, Florida, USC, Texas, and Texas Tech, as shown in Table 11.2. Every undefeated team has a legitimate claim to the #1 ranking, but in 2008, Utah was the only such team. Four other teams could also appear atop a perfect ranking, despite one or more losses. Florida's only defeat was the previously-mentioned upset by Mississippi, so there is no team which must be ranked ahead of Florida. Oklahoma, Texas, and Texas Tech were part of a cyclic triad, so any of these could be ranked ahead of the other two. Having no other losses, Texas could then be ranked #1. Texas Tech had one other loss, to Mississippi. Even two wins over highly-ranked teams would not allow Mississippi to be placed above either Texas Tech or Florida, thus Texas Tech could land in the top spot. However, Oklahoma must be ranked below Florida, because of a loss to Florida. Finally, Southern Cal could be #1, with a record blemished only by an upset loss to Oregon State.

There is a wide range of potential rankings for some teams. No team appears in the top five of every perfect ranking, and only three teams — Florida,

USC, and Oklahoma — are even guaranteed a spot in the top ten. Despite being the only undefeated team, Utah can be ranked as low as #13. Texas could fall as far as #21, and Alabama (ranked 6th by consensus of the BCS computer rankings, see Appendix A) could be altogether removed from the top 25. It is also possible to move teams higher than their results seem to warrant. For example, there is a perfect ranking in which the top four teams (in descending order) are Florida, Florida State (7-4), Virginia Tech (9-4), and Maryland (7-5). The latter three, all from the relatively weak Atlantic Coast Conference, finished the season ranked 18th, 16th, and 35th, respectively, by consensus. The much stronger Big 12 Conference provides another example of how whole conferences can be over-ranked or under-ranked, due to the strong interconnectedness of teams within a conference. Table 11.3 lists the top 25 for two perfect rankings, with very different results for the Big 12. In the ranking on the left, eight of the top sixteen teams are from the Big 12, including 4-7 Colorado at #16. In the ranking on the right, the Big 12 has just one of the top 19 teams.

Just as there are can be questionable outcomes at the top of a perfect ranking, the same is true at the bottom. Traditional power Michigan had a miserable season by its standards, winning just three games. However, two victories came against bowl-bound teams, Wisconsin and Minnesota. Because these upset wins are largely ignored, the Wolverines could be ranked as low as #119 out of 120 FBS teams, even though there were 28 other teams which won three or fewer games. Middle Tennessee State won five games, including a victory over a Maryland team that eventually won a bowl game, yet MTSU can be ranked as low as #117. On the other hand, Washington, one of only three winless FBS teams, can be ranked in the upper half at #57. For comparison, the consensus rankings of Michigan, MTSU, and Washington are # 94, #97, and #115, respectively. While less relevant for applications involving the BCS, these results demonstrate weaknesses of a ranking system based solely on minimizing violations.

In general, MinV penalizes a defeat to a far-inferior team less than a loss to a similarly-ranked or slightly inferior opponent. Counterintuitively, a contending team with a single loss is ranked higher if that loss came against a weak opponent, rather than a good one. On the other hand, upsetting a far-superior team may not result in a ranking improvement, while beating a slightly-superior team will generally be rewarded. At season's end, Mississippi was ranked in the top 20 of both major polls and all six BCS computer rankings, largely on the strength of twin upsets of Florida and Texas Tech, but minimizing violations requires that Ole Miss be ranked below three

Table 11.3. Two perfect rankings, with the one on the left favorable for the Big 12 conference, and the one on the right unfavorable for the Big 12. In both lists, Big 12 teams are shown in bold type. Listed after each team's record is its consensus rank among the six BCS computer rankings.

1) Texas Tech (9-2, #8)	1) Florida (12-1, #2)
2) Texas (12-1, #3)	2) Southern Cal (12-1, #4)
3) Florida (12-1, #2)	3) Utah (12-0, #1)
4) Oklahoma (11-2, #5)	4) Alabama (12-2, #6)
5) Southern Cal (12-1, #4)	5) Penn State (10-2, #11)
6) Utah (12-0, #1)	6) Florida State (7-4, #18)
7) TCU (10-2, #7)	7) Virginia Tech (9-4, #16)
8) Boise State (11-1, #10)	**8) Oklahoma** (11-2, #5)
9) Oregon (10-3, #12)	9) TCU (10-2 #7)
10) Oklahoma State (8-4, #22)	10) Boise State (11-1, #10)
11) Missouri (9-4, #19)	11) Oregon (10-3, #12)
12) Florida State (7-4, #18)	12) Oregon State (9-4, #15)
13) Virginia Tech (9-4, #16)	13) Cincinnati (10-3, #17)
14) Nebraska (9-4, #20)	14) Pitt (9-4, #24)
15) Kansas (7-5, #31)	15) Arizona (8-5, #38)
16) Colorado (4-7, #60)	16) Maryland (7-5, #35)
17) Alabama (12-2, # 6)	17) California (9-4, #21)
18) Penn State (10-2, # 11)	18) Brigham Young (9-3, #28)
19) Oregon State (9-4, # 15)	19) Air Force (7-5, #49)
20) Ohio State (9-3, #13)	**20) Texas Tech** (9-2, #8)
21) Arizona (8-5, # 38)	**21) Texas** (12-1, #3)
22) Maryland (7-5, # 35)	22) Ohio State (9-3, #13)
23) California (9-4, # 21)	**23) Oklahoma State** (8-4, #22)
24) Cincinnati (10-3, # 17)	**24) Missouri** (9-4, #19)
25) Pitt (9-4, # 24)	**25) Nebraska** (9-4, #20)

mediocre teams to which they lost. These principles are in opposition to those of most other ranking methods, in which every upset carries weight, a loss to a weak opponent is more damaging than a loss to a strong one, and defeating a top team is substantially rewarded.

The existence of multiple perfect rankings was an issue known to Coleman at the time of his 2005 paper [9]. Logically, the next step in extending a violation-minimizing method is to add additional criteria to narrow down the possible outcomes. Beginning with the 2006 season, Coleman has used victory margins to determine which on-field results should be violated when there is a choice, for example, in the Texas–Texas Tech–Oklahoma cyclic triad. In the games among those three teams, the smallest margin of victory was Texas Tech's six-point win over Texas, so this result is violated. The alternatives would have been to violate either Texas' ten-point win over Oklahoma or Oklahoma's 44-point win over Texas Tech. Thus, Texas is rated above Oklahoma, which is rated above Texas Tech. While the use of margin-of-victory in rankings is contentious (and prohibited in the BCS computer rankings), Coleman's choice is logical. For the 2008 season, Coleman determined that, with the minimum 69 violations, the smallest possible sum of the margins of violated games is 573 points [8]. This does not make the ranking unique (of the five possible #1 teams, only Texas Tech is eliminated from potentially topping the ranking). To narrow further the possibilities toward uniqueness, the MinV rankings use the Sagarin rankings [19] as a final criterion, attempting to most closely match Sagarin's rankings while maintaining the minimum number of violations and total margin of violations. This is an unsatisfactory solution, as it is preferable for ranking systems to each stand on their own, independent of others. A version of the Sagarin ratings is already used in the BCS formula, so adding the MinV ratings to the BCS would overweight Sagarin's ratings.

GridMap, also designed to minimize violations, takes another approach to resolving loops. It utilizes a different algorithm (which is not fully disclosed), focusing on intertwined loops, in which one or more games are a part of multiple loops. Each violation is listed, along with the sequence of games in the associated loop (since every violation involves a loop). In Massey's 2008 comparison of ranking systems, GridMap finished second only to MinV in violation percentage [15]. The approach appears to be less computationally intensive than Coleman's method, and no external ranking is used. The result is generally not a retrodictively perfect ranking, and many unbroken ties are included in the final results. In 2008, GridMap had Utah at #1, followed by a three-way tie between Florida, Texas, and Southern Cal, and then a four-way tie at #5 between Oklahoma, Alabama, Penn State, and Ball State. It is notable that Ball State, which achieved an 11-2 record against a weak schedule, was not ranked in the final top 25 by MinV, either major poll, or any of the BCS computer rankings. Also, the GridMap algorithm ranks winless Washington (consensus #115) ahead of 10-3 Tulsa (consensus #40). This result indicates that the method used by GridMap, like early versions of MinV, may lead to questionable outcomes. These occur, at least in part, because GridMap's method ignores any violated game result for ranking purposes, so (as with MinV), the most surprising upsets make little impact on the GridMap rankings.

The rankings resulting from Coleman's work may not be suitable for inclusion in the BCS rankings at this time, yet they remain useful because they are based on sound logical principles and provide contrarian views to those of other ranking systems. The huge retrodictive advantage of MinV, as compared to previous methods, shows that there is a great deal of room for improvement among other types of computational methods in accurately reflecting prior on-field results. Perhaps new ranking systems will emerge, combining near-perfect retrodiction with a high degree of predictive accuracy, in a way that meaningfully accounts for upsets. Standing on its own, the MinV algorithm is a significant advance, using the power of high-speed computers to achieve the highest possible retrodictive success.

Acknowledgments

Thanks to Jay Coleman and Warren Repole for providing historical game scores.

References

[1] J. Anderson, The Anderson & Hester College Football Computer Rankings, *Anderson Sports*.
www.andersonsports.com/football/ACF_frnk.html

[2] ——, AP Football Poll Archive, *AP Poll Archive*.
www.appollarchive.com/football/index.cfm

[3] T. Beck, College Football Ratings PT Awards, *The Prediction Tracker*, accessed on 16 July 2009. tbeck.freeshell.org/fb/awards2008.html

[4] ——, Computer Rating System Prediction Results for College Football, *The Prediction Tracker*, accessed on 16 July 2009.
www.thepredictiontracker.com/ncaaresults.php?type=1&year=08

[5] R. Billingsley, Billingsley Report on Major College Football, *College Football Research Center*, www.cfrc.com

[6] ——, BCS Chronology, 25 June 2009, *Bowl Championship Series*, accessed on 11 August 2009. www.bcsfootball.org/bcsfb/history

[7] ——, BCS Computer Rankings, 11 Nov. 2008, *Bowl Championship Series*, accessed on 21 July 2009. www.bcsfootball.org/bcsfb/rankings

[8] J. Coleman, Minimum Violations (MinV) College Football Ranking, 10 Jan. 2009, *University of North Florida*, accessed on 8 July 2009.
www.unf.edu/~jcoleman/minv.htm

[9] ——, Minimizing Game Score Violations in College Football Rankings, *Interfaces* 35 (2005) 483-496.

[10] W. N. Colley, Colley's Bias Free Matrix Rankings, *Colley Matrix*.
www.colleyrankings.com

[11] ——, Colley's Bias Free College Football Ranking Method, 30 May 2002, *Colley Matrix*, accessed on 25 Aug. 2009. www.colleyrankings.com/method.html

[12] M. R. Garey and D. S. Johnson, *Computers and Intractability: A Guide to the Theory of NP-Completeness*, W. H. Freeman and Co., New York, 1979.

[13] ——, GridMap post-Bowls, *GridMap of College Football*, accessed on 15 July 2009. Unable to access as of May 2010. www.gridmaponline.com

[14] ——, Harris Interactive College Football Poll, *Harris Interactive*. www.harrisinteractive.com/news/bcspoll.asp

[15] K. Massey, College Football Ranking Comparison, 23 Mar. 2009, *Massey Ratings*, accessed on 22 July 2009. www.mratings.com/cf/compare.htm

[16] ——, College Football Rankings, *Massey Ratings*. www.masseyratings.com/rate.php?lg=cf

[17] ——, Massey Ratings Description, 15 Aug. 2000, *Massey Ratings*, accessed on 28 July 2009. www.mratings.com/theory/massey.htm

[18] W. Repole, Historical Scores, 31 May 2009, *The Sunshine Forecast*, accessed on 23 June 2009. www.repole.com/sun4cast/scores2009.html

[19] J. Sagarin, Jeff Sagarin NCAA Football Ratings, 9 Jan. 2009, *USA Today*, accessed on 22 July 2009. www.usatoday.com/sports/sagarin/fbt08.htm

[20] ——, Threads, Loops, and Upsets, *GridMap of College Football*, accessed on 15 July 2009. Unable to access as of May 2010. www.gridmaponline.com/#/threads-loops-and-upsets/4525582180

[21] ——, USA Today Coaches' Poll, *USATODAY.com*. www.usatoday.com/sports/college/football/usatpoll.htm

[22] D. L. Wilson, American College Football — Rankings, *Nutshell Sports Ratings*, accessed on 22 July 2009. www.nutshellsports.com/wilson/popup.html

[23] P. Wolfe, Ratings of all NCAA and NAIA Teams. prwolfe.bol.ucla.edu/cfootball/home.htm

About the Author

R. Drew Pasteur is an Assistant Professor of Mathematics at The College of Wooster. Before pursuing graduate study at North Carolina State University, he was a math teacher and head athletic trainer at Fuquay-Varina High School, in Fuquay-Varina, NC. It was during this time that Drew became interested in mathematical ranking algorithms and developed the first version of his Fantastic 50 high school football rankings. His other research area is in mathematical biology, applying dynamical systems to human physiology. Outside of mathematics, Drew enjoys singing, running, and spending time with his wife and son.

Appendix
Compilation of various rankings

Retrodictive Rankings: MinV [8], GridMap [13]
BCS Computer Rankings: Sagarin [19], Anderson/Hester [1], Billingsley [5], Colley [10], Massey [16], Wolfe [23]
Major Polls: Associated Press [2], USA Today Coaches' [21]

Consensus rankings were determined by dropping the highest and lowest rank for each team among the six BCS computer rankings, then averaging the remaining four. Ties were broken by using the two dropped rankings.

Consensus ranking among BCS computer rankings	Retrodictive		BCS computer rankings						Polls	
	MinV	GMap	Sag	And	Bil	Col	Mas	Wol	AP	USA
1) Utah (12-0)	5	1	1	1	3	3	1	1	2	4
2) Florida (12-1)	1	2-T	2	2	1	1	2	2	1	1
3) Texas (12-1)	3	2-T	3	3	4	2	4	3	4	3
4) USC (12-1)	2	2-T	5	5	2	4	5	4	3	2
5) Oklahoma (11-2)	4	5-T	4	4	5	5	3	5	5	5
6) Alabama (12-2)	6	5-T	6	6	6	6	6	7	6	6
7) TCU (10-2)	7	9-T	7	7	8	11	8	6	7	7
8) Texas Tech (9-2)	9	9-T	9	8	7	7	7	9	12	12
9) Georgia (9-3)	29-T	17-T	8	11	12	10	9	8	13	10
10) Boise State (11-1)	10	12-T	15	9	9	9	10	12	11	13
11) Penn State (10-2)	8	5-T	13	10	10	8	11	11	8	8
12) Oregon (10-3)	11	14-T	11	13	14	14	12	10	10	9
13) Ohio State (9-3)	13	9-T	17	12	11	12	17	15	9	11
14) Mississippi (8-4)	59-T	66-T	10	20	13	19	13	13	14	15
15) Oregon State (9-4)	12	17-T	12	16	19	16	14	14	18	19
16) Virginia Tech (9-4)	15	12-T	14	18	20	15	15	16	15	14
17) Cincinnati (10-3)	22-T	14-T	19	14	22	13	19	17	17	17
18) Florida State (7-4)	14	21-T	16	21	26	20	16	19	21	23
19) Missouri (9-4)	17	21-T	22	15	21	18	22	20	19	16
20) Nebraska (9-4)	18	25-T	21	19	25	22	20	22		
21) California (9-4)	21	28-T	18	24	24	21	23	18		25
22) Oklahoma State (8-4)	16	17-T	24	22	15	25	21	21	16	18
23) Georgia Tech (7-4)	28	14-T	23	26	18	24	18	28	22	22
24) Pittsburgh (9-4)	24-T	21-T	27	17	29	17	28	25		
25) Michigan State (9-4)	33	31-T	31	23	23	23	31	27	24	24
26) West Virginia (8-4)	26	33-T	25	27	28	27	29	24	23	
27) LSU (7-5)	64	75-T	20	34	17	38	30	23		
28) BYU (9-3)	32	25-T	34	25	16	26	26	31	25	21
29) Wake Forest (8-5)	52-T	43-T	26	28	47	31	25	29		
30) Boston College (8-5)	31	17-T	28	29	38	29	24	32		
31) Kansas (7-5)	22-T	28-T	32	30	34	34	32	30		
32) North Carolina (7-5)	27	40	33	31	41	33	27	37		
33) Iowa (8-4)	35	35	37	32	32	32	38	33	20	20
34) Rice (10-3)	59-T	66-T	39	33	39	30	40	26		
35) Maryland (7-5)	20	25-T	30	37	35	37	35	35		

Consensus rankings	MinV	GMap	Sag	And	Bil	Col	Mas	Wol	AP	USA
36) Vanderbilt (7-6)	57-T	62-T	29	38	36	43	37	34		
37) Rutgers (7-5)	45	55-T	38	40	54	40	39	39		
38) Arizona (8-5)	19	21-T	35	48	27	48	44	36		
39) Connecticut (7-5)	47-T	59-T	42	39	43	39	43	38		
40) Tulsa (10-3)	68	80-T	50	42	30	35	48	40		
41) Northwestern (8-4)	34	33-T	48	36	33	36	46	47		
42) Miami-FL (6-6)	50-T	41-T	36	44	60	45	34	42		
43) South Carolina (6-6)	36	36-T	40	41	45	44	36	43		
44) Ball State (11-2)	69	5-T	62	35	31	28	54	51		
45) Clemson (5-6)	29-T	28-T	41	46	40	46	33	48		
46) South Florida (7-5)	46	57-T	45	45	37	41	45	45		
47) East Carolina (9-5)	67	75-T	46	47	50	42	50	41		
48) Kentucky (6-6)	59-T	66-T	44	50	49	52	47	44		
49) Air Force (7-5)	39-T	51	57	43	51	47	49	53		
50) Houston (7-5)	62-T	71-T	52	54	48	51	58	46		
51) Virginia (4-7)	57-T	62-T	43	58	52	57	41	54		
52) Colorado State (6-6)	41-T	31-T	53	49	77	53	52	49		
53) Notre Dame (7-6)	37	41-T	51	52	59	54	56	52		
54) Navy (7-5)	70	49-T	54	51	62	50	53	56		
55) NC State (5-7)	47-T	38-T	47	55	58	56	42	57		
56) Arkansas (4-7)	62-T	71-T	49	57	57	62	51	50		
57) Western Michigan (8-4)	77	71-T	64	53	42	49	63	55		
58) Wisconsin (6-6)	41-T	36-T	60	56	44	55	60	60		
59) Stanford (5-7)	54-T	59-T	56	67	53	67	59	58		
60) Colorado (4-7)	24-T	52	58	59	66	68	57	59		
61) Auburn (4-7)	65	80-T	59	64	55	71	62	61		
62) Tennessee (5-7)	54-T	59-T	61	61	63	70	61	64		
63) Louisiana Tech (7-5)	41-T	49-T	66	63	61	61	68	65		
64) Nevada (6-6)	39-T	46-T	65	60	68	59	65	66		
65) Minnesota (6-6)	73	38-T	71	62	64	63	71	69		
66) Arizona State (4-7)	50-T	55-T	63	74	46	73	64	74		
67) Duke (3-8)	54-T	46-T	55	78	75	76	55	70		
68) Hawaii (6-7)	38	43-T	74	68	69	65	67	78		
69) Buffalo (8-6)	78-T	75-T	73	66	86	60	76	63		
70) Southern Miss (7-6)	66	99	67	75	71	69	77	62		
71) Fresno State (7-6)	44	53-T	75	69	83	66	70	71		
72) Louisville (4-7)	89	95-T	69	76	70	75	72	72		
73) Troy (7-5)	72	100	78	71	74	64	80	67		
74) Mississippi State (3-8)	81	85-T	68	83	67	86	69	76		

Consensus rankings	MinV	GMap	Sag	And	Bil	Col	Mas	Wol	AP	USA
75) Illinois (4-7)	78-T	75-T	77	73	65	74	75	75		
76) Baylor (3-8)	71	62-T	70	72	78	79	66	80		
77) Central Michigan (7-5)	76	46-T	86	65	80	58	82	73		
78) Kansas State (4-7)	90	97-T	76	77	76	80	74	77		
79) UCLA (4-8)	52-T	57-T	72	84	56	81	73	81		
80) Florida Atlantic (7-6)	75	101	80	80	79	72	86	68		
81) UNLV (5-7)	47-T	53-T	79	70	89	78	78	83		
82) Texas A&M (4-8)	105-T	108-T	81	79	81	85	79	88		
83) San Jose St (5-6)	91	62-T	83	81	82	77	84	82		
84) New Mexico (4-8)	80	85-T	82	82	88	89	81	84		
85) Purdue (3-8)	74	43-T	85	85	73	84	83	87		
86) UTEP (5-7)	96-T	95-T	87	86	87	83	87	79		
87) Arkansas State (5-6)	103-T	106-T	90	90	91	87	95	85		
88) Memphis (5-7)	101-T	97-T	89	89	97	88	88	92		
89) Bowling Green (6-6)	83	80-T	92	87	99	82	92	90		
90) Syracuse (2-9)	88	93-T	84	91	100	94	85	91		
91) Louisiana-Lafayette (6-6)	98-T	102-T	91	93	90	90	94	86		
92) Marshall (3-8)	100	85-T	88	95	95	92	89	89		
93) Wyoming (3-8)	85	88	95	88	84	96	90	99		
94) Michigan (3-9)	105-T	108-T	94	94	72	98	91	101		
95) Northern Illinois (5-7)	82	75-T	100	92	101	91	96	96		
96) Temple (5-7)	84	80-T	99	96	102	93	97	94		
97) MTSU (5-7)	107	108-T	97	100	94	99	98	93		
98) Akron (5-7)	87	91-T	101	98	106	95	100	95		
99) Florida International (5-7)	101-T	104-T	102	99	98	97	103	97		
100) Utah State (3-9)	92-T	66-T	98	97	110	100	93	102		
101) UCF (3-8)	98-T	104-T	93	102	107	101	99	100		
102) UAB (3-8)	96-T	102-T	96	103	104	103	101	98		
103) Indiana (2-9)	109	80-T	104	101	85	102	102	105		
104) Ohio U. (3-8)	86	89-T	110	104	103	104	108	103		
105) Washington State (1-11)	112	66-T	103	109	92	108	106	112		
106) Iowa State (1-10)	108	111-T	106	105	96	113	107	111		
107) Louisiana-Monroe (3-8)	115	116-T	109	108	108	105	110	104		
108) New Mexico State (2-9)	95	93-T	108	107	111	107	109	107		
109) Kent State (3-8)	111	113-T	112	110	105	106	111	106		
110) Army (3-8)	110	111-T	105	111	114	109	104	109		
111) San Diego State (2-9)	92-T	89-T	107	106	113	111	105	113		
112) Toledo (3-9)	103-T	106-T	111	113	109	112	112	108		
113) Eastern Michigan (2-9)	116	113-T	115	112	115	110	113	110		

Consensus rankings	MinV	GMap	Sag	And	Bil	Col	Mas	Wol	AP	USA
114) Tulane (2-10)	114	113-T	114	114	116	114	114	114		
115) Washington (0-12)	113	71-T	113	116	93	115	116	119		
116) Idaho (1-10)	94	91-T	116	115	118	116	115	115		
117) Miami-OH (1-10)	118	116-T	118	118	112	119	117	116		
118) Southern Methodist (0-11)	117	116-T	117	117	117	118	118	118		
119) North Texas (1-11)	119	119	119	119	119	120	119	117		
120) Western Kentucky (0-10)	120	120	120	120*	120	117	120	120		

Notes:

- The AP and USA Today polls involve voting only for the top 25 teams, while computer rankings generally give a complete ranking of the 120 FBS teams.

- The final Anderson/Hester rankings listed only 119 teams. Because winless Western Kentucky was omitted, we assume it to be #120 in this ranking.

- The final Harris Interactive Poll voting occurs prior to the bowl games, so we have not included it.

- Only games against FBS teams are included in the teams' records.

Part IV

Golf

CHAPTER 12

The Science of a Drive

Douglas N. Arnold

Abstract

Golf provides numerous examples of common physical phenomena which can be elucidated through mathematics. This notes provides a simple introduction to mathematical modeling in golf, by briefly describing a few of the many ways mathematics can be used to understand or improve the golf drive. First we describe the double-pendulum model of a golf swing, which is a simple but useful model of the mechanical system consisting of the golfer and the golf club, used to accelerate the club head. Second we consider the basic mechanics of the energy and momentum transfer which takes place when the club head impacts the golf ball. Finally we describe the three basic forces—gravity, drag, and lift—which determine the ball's trajectory after it is struck by the club.

"Math and science are everywhere." With those words, championship golfer Phil Mickelson began a public service television advertisement produced by ExxonMobil and premiered during the 2007 broadcast of the Masters Golf Tournament. I had the privilege to serve as the mathematical consultant for the ad and for the accompanying website, *The Science of a Drive*, from which the title of this article is taken. Figure 12.1 displays a still frame taken from the advertisement and another taken from the website.

The golf drive does indeed provide numerous examples of the ways mathematics elucidates common physical phenomena. Many aspects of it can be illuminated or improved through mathematical modeling and analysis of the mechanical processes entering into the game. Here I present a few simple examples collected during my consulting work. Specifically I briefly discuss three applications of mathematical modeling to fundamental mechanical processes in the golf drive: the double-pendulum model of a golf swing, transfer of energy and momentum in the club head/ball impact, and drag and lift in the flight of the golf ball.

Figure 12.1. Frames from the television advertisement and the website.

These examples just scratch the surface of the subject. Indeed, there is a large literature on the subject of mathematics and mechanics of golf. See, for example, the survey [5] which discusses several aspects:

- models of the golf swing,

- the physics of the golf club and ball,

- the impact of the club head and the golf ball,

- golf ball aerodynamics,

- the run of the golf ball on turf.

12.1   The double-pendulum approximation of the swing

When a golfer swings for a long drive, the goal is to accelerate the club head so that it impacts the ball at just the right point, going in just the right direction, and moving as quickly as possible. To do so, the golfer exerts force with his or her arms on the shaft of the club, which in turn exerts force on the club head. This situation may be approximated as a double pendulum as depicted in Figure 12.2. The arms, pivoting at the shoulders, roughly behave as a pendulum, and the hands, grip, and shaft, pivoting at the wrists, behave as a second pendulum attached at the end of the first. For a well-timed drive, at the moment of impact the upper pendulum—the arms—is swinging very rapidly about its pivot point, and, at the same moment, the club is swinging very rapidly around its pivot point. These movements combine to accelerate the club head to speeds as high as 120 miles per hour.

Of course the double pendulum model is a crude approximation of the complex mechanism formed by the body and the club during a swing. The model can be refined in many ways, for example by taking into account the movement of the shoulders (and so of the pivot point of the upper pendulum)

Figure 12.2. The double pendulum model of a golf swing.

[7], the flexing of the club shaft [3], and the three-dimensional aspects of the motion [4].

12.2 The impact of the club head and the ball

The velocity of the club head, together with its mass, determine its kinetic energy and momentum. As the swing progresses, the golfer applies more and more force to the club head causing it to accelerate and so increase its speed. Therefore its momentum and energy increase. Upon impact, some of this energy and momentum is transferred to the ball. To determine the speed of the ball as it leaves the tee, we use conservation of both energy and momentum. Let m_{club} and m_{ball} denote the mass of the club and the ball, respectively. Let V_{club} and v_{ball} denote their speeds right after impact, and let v_{club} denote the speed of the club head just before impact. (Of course the speed of the ball just before impact is zero.) Since $E = mv^2/2$, conservation of energy tells us that

$$\frac{1}{2}m_{club}v_{club}^2 = \frac{1}{2}m_{club}V_{club}^2 + \frac{1}{2}m_{ball}v_{ball}^2,$$

while conservation of momentum tells us that

$$m_{club}v_{club} = m_{club}V_{club} + m_{ball}v_{ball}.$$

The solution to these equations is easily found:

$$V_{club} = v_{club}\frac{m_{club} - m_{ball}}{m_{club} + m_{ball}}, \qquad v_{ball} = v_{club}\frac{2m_{club}}{m_{club} + m_{ball}}$$

$$= v_{club}\frac{2}{1 + m_{ball}/m_{club}}.$$

Thus the ratio of the ball speed to the speed of the club head before impact is $2/(1 + r)$ where r is the ratio of the mass of the ball to the the mass of the club head. Notice that, no matter how small the ratio of masses, the ball speed will always be less than twice the club head speed. For instance, if $v_{\text{club}} = 54.0$ meters per second (about 120 miles per hour), $m_{\text{club}} = 0.195$ kilograms, and $m_{\text{ball}} = 0.0459$ kilograms, then v_{ball} is about 87.4 meters per second or just about 195 miles per hour.

In reality, not all of the kinetic energy lost by the club head during impact is converted into kinetic energy of the ball. That is, the impact is not perfectly elastic. Some energy is lost to heat and damage to the ball. In this case, the ball launch speed is given by

$$v_{\text{ball}} = \frac{(1 + c_R)v_{\text{club}}}{1 + m_{\text{ball}}/m_{\text{club}}} \tag{12.1}$$

where c_R is called the *coefficient of restitution*. For an elastic collision, $c_R = 1$, but in reality it is somewhat smaller. Using a typical value of $c_R = 0.78$, we obtain a launch velocity $v_{\text{ball}} = 77.8$ meters per second, or about 175 miles per hour. Even to the nonspecialist, formula (12.1) conveys a sense that math impinges on golf, and it was prominently displayed in the television advertisement (see Figure 12.1).

The period of contact of the club head with the ball is about one two-thousandth of a second. During this time the center of mass of the ball has barely moved, but the ball is bent way out of shape. A significant portion of the kinetic energy has been converted into potential energy stored in the deformed ball. Essentially, the ball is like a compressed spring. See Figure 12.3. When the ball takes off from the tee, it returns to a spherical shape, releasing the spring, and most of this potential energy is converted back into kinetic energy. Detailed analyses of the club head/ball interaction can be made through a full 3-dimensional finite element analysis [2] or via simplified 1- or 2-dimensional models [1].

12.3   The ball's flight

Once the ball is in flight, its trajectory is completely determined by its launch velocity and launch angle and the forces acting on it. The most important of these forces is, of course, the force of gravity, which is accelerating the ball back down towards the ground at 9.8 meters per second per second. But the forces exerted on the ball by the air it is passing through are important as well. To clarify this, we choose a coordinate system with one axis aligned with the direction of flight of the ball and the others perpendicular to it. Then the forces exerted by the atmosphere on the ball are decomposed into the *drag*, which is a force impeding the ball in its forward motion, and the

Figure 12.3. Golf ball under compression from impact of club on left.

lift which helps the ball fight gravity, and stay aloft longer (Figure 12.4). Drag is the same force you feel pushing on your arm if you stick it out of the window of a moving car. Lift is a consequence of the back spin of the ball, which speeds the air passing over the top of the ball and slows the air passing under it. By Bernoulli's principle the result is lower pressure above and therefore an upward force on the ball.

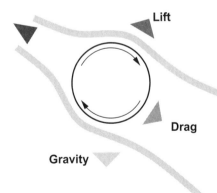

Figure 12.4. Forces acting on a golf ball during its flight.

Drag and lift are very much affected by how the air interacts with the surface of the ball. The dimples on a golf ball are there primarily to decrease drag and increase lift. Proper dimpling of a golf ball induces turbulence in the boundary layer, delaying the point at which the flow past the ball separates from the surface, and resulting in a ball which can carry nearly twice as far as a smooth ball would with the same swing. Based on aerodynamic and manufacturing considerations, a great many dimple designs have been manufactured, leading to an elaborate crystallography of golf balls [6].

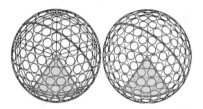

Figure 12.5. Two golf ball dimple patterns with icosahedral symmetry.

Mathematics Awareness Month 2010, with the theme Mathematics and Sports, provides mathematicians with another opportunity to get out the vitally important message that mathematics can be found everywhere in the physical world and human activity. In this note, I have discussed briefly a few of the ways in which mathematics relates to golf. All of these could be, and in fact have been, the subject of extended studies, because they enable us not only to better understand, but also to optimize, the performance of a golfer and golf equipment.

References

[1] Alistair J. Cochran, *Development and use of one-dimensional models of a golf ball*, Journal of Sports Sciences **20** (2002), 635–641.

[2] Takuzo Iwatsubo, Shozo Kawamura, Kazuyoshi Miyamoto, and Tetsuo Yamaguchi, *Numerical analysis of golf club head and ball at various impact points*, Sports Engineering **3** (2000), 195–204.

[3] Sasho J. MacKenzie, *Understanding the role of shaft stiffness in the golf swing*, PhD dissertation, University of Saskatchewan, 2005.

[4] Steven M. Nesbit, *A three dimensional kinematic and kinetic study of the golf swing*, Journal of Sports Science and Medicine **4** (2005), 499–519.

[5] A. Raymond Penner, *The physics of golf*, Reports on Progress in Physics **66** (2003), 131–171.

[6] Ian Stewart, *Crystallography of a golf ball*, Scientific American **276** (1997), 96–98.

[7] A. B. Turner and N. J. Hills, *A three-link mathematical model of the golf swing*, Science and golf, III: proceedings of the 1998 World Scientific Congress of Golf, Human Kinetics, Champaign, pp. 3–12.

About the Author

Douglas Arnold is McKnight Presidential Professor of Mathematics at the University of Minnesota and President of the Society for Industrial and Applied Mathematics (SIAM). His research interests include numerical analysis, partial differential equations, mechanics, and in particular, the interplay between these fields. He initiated the finite element exterior calculus, a new approach to the stability of finite element methods based on geometric and topological considerations, and its development is a major direction of his current research work. Arnold was a plenary speaker at the ICM in Beijing in 2002 and a Guggenheim Fellow in 2008-2009, and delivered an AMS–MAA invited address at the Joint Math Meetings in 2009. He is a foreign member of the Norwegian Academy of Science and Letters and a SIAM Fellow. Arnold also takes a strong interest in the teaching and communication of mathematics, and is the coauthor of the award-winning video *Möbius Transformations Revealed,* which became a runaway hit on YouTube, with 1.7 million views. He has also served on numerous mathematical advisory committees, including the Program Committee of the International Congress of Mathematicians, the U.S. National Committee for Mathematics, and the scientific boards of numerous mathematical institutes in the U.S. and abroad. From 2001 through 2008 he served as director of the Institute for Mathematics and its Applications (IMA), which, during that period, grew to become the largest mathematics research investment in the history of the National Science Foundation.

CHAPTER 13

Is Tiger Woods a Winner?

Scott M. Berry

Abstract

Tiger Woods has an amazing record of winning golf tournaments. He has gained the persona of a player who is a winner, a player who when near the lead or in the lead can do whatever it takes to win. In this paper I investigate whether in fact, he is a winner. A mathematical model is created for the ability of Tiger Woods, and all PGA Tour golfers to play 18 holes of tournament golf. The career of Tiger Woods is replayed using the mathematical model for all golfers and the results are consistent with Tiger Woods' actual career. Therefore the mathematical model, which does not give Woods any additional ability to win, would result in essentially the same career. Woods has not needed any additional ability to win—only his pure golfing ability. The ramifications of this result are that there is no evidence that Woods is in fact a "winner"—instead he is just a much better golfer than everyone else.

13.1   Introduction

Tiger Woods is one of those rare athletes who accomplish feats in their sport that are freakish. In this small group are guys such as Babe Ruth, Wayne Gretzky, Wilt Chamberlain, Barry Bonds, and Jack Nicklaus. Woods dominates a sport where the population of players are all very good–and very tightly bundled in their ability. To win one tournament, beating 100+ of these players is incredibly difficult. To average one tournament victory a year for 10 years is a Hall of Fame type accomplishment. Tiger Woods has won 71 of 253 (28.1%) official PGA Tour tournaments around the world in 14 years on the PGA Tour. The PGA Tour record is 82 career wins by Sam Snead in 30 years on the PGA Tour. In 2009 Woods played 17 events and won 6 of them, finishing in the top 10 in an astounding 14 of them. Despite the 6 wins in 17 events, some classify his season as disappointing because he didn't win any of the four tournaments that are labeled "majors." The Masters, United States Open, The Open Championship, and the PGA Championship are the four so-called majors in golf. These tournaments have extra importance in professional golf. They are also different from typical tournaments because they typically provide the best fields—the quality and depth of the players participating. Woods has won 14 career Majors in 49 played as a professional. The record is 18, held by Jack Nicklaus, who competed in 163 majors (he also finished second 19 times). Woods holds the record for the largest margin of victory in each major—12 in the Masters, 15 in the US Open, 8 in The Open Championship, and 5 in the PGA Championship.

Tiger has accomplished a huge amount despite being only 33 years old. Professional golfers are typically competitive into their mid-forties (Tom Watson nearly won The Open Championship in 2009 at the age of 59!). Barring any catastrophic injuries or strange happenings, Woods should shatter all career win records, including most tour victories and most major wins. Is it then ludicrous to ask the question whether Tiger Woods is a "winner." First, let me define what I mean by a "winner." I think Tiger Woods is the greatest golfer, if not the greatest athletic performer of all-time. But, by winner, I want to know if he has an ability to win golf tournaments beyond his skill level. Woods has developed the persona of a champion, that his presence on the leaderboard—being near the lead late in the tournament—causes other players to perform worse. Thus, Tiger has an innate ability to win. Nicklaus, with whom Woods is most compared, won 18 majors and finished second 19 times. Woods has won 14 majors, but finished second "only" 6 times. The thought is that if Woods were not a winner, he should have a distribution of firsts and seconds similar to Nicklaus.

Many commentators claim that Woods' best attribute is his "mind." That he is able to do whatever it takes to win. If this is true, that Woods has an innate ability to win golf tournaments, then he should win more than his raw ability. There are many different "effects" that are explored in sports, including the so called "hot hand effect" or the ability to hit in the clutch. In this paper I explore whether Woods wins more than his raw golfing ability would determine he wins—whether Tiger Woods is a "winner."

In this paper I create a mathematical model for Tiger Woods. In this process I also create a mathematical model for all professional golfers. In this model each player will have an intrinsic ability to play golf. In this model Woods—I'll refer to him as "RoboTiger"—will not have the ability to win beyond his golfing skill. Each player will record a score on each round, independently of what every other player is scoring. RoboTiger will not perform differently when he is near the lead, and likewise none of his competitors will perform worse when RoboTiger is near the lead. RoboTiger will not be a winner. The mathematical model of Woods will rank him as the best player—therefore RoboTiger will win tournaments, but he will not be an innate winner beyond his mathematical golfing skills. I simulate ten thousand replays of Woods' career, using RoboTiger and all the mathematical golfers to compare what RoboTiger does relative to Woods' actual career record. Should RoboTiger win less than the real Tiger, then this is evidence that the real Tiger may be an innate "winner."

13.2   Data Availability and Mathematical Modeling

Every year there are approximately 160 players who qualify for full tour status on the Professional Golfers of America (PGA) Tour. There are several hundred more that play fewer tournaments with partial tour status. A number of foreign players will play a small number of PGA events in addition to their native professional tours. Each season consists of approximately 45 official PGA Tour tournaments. For many of the important tournaments the best players from around the world participate. For some of the perceived lesser events many of the top players do not play. A typical tournament will have 144 players for the first two rounds, then a cut occurs where the field is narrowed to the top 60 players. I downloaded the round-by-round scores for each of the tournaments on the PGA Tour for 1997 (Woods' first full season) through 2004. Over this eight-year period there were 2004 different players participating in the 352 tournaments (match play and Stableford scoring tournaments were not included). There were 147,154 rounds played, with 10,509,253 shots taken. The best scoring average for any player was 68.75 for Chris Downes. Woods had the fourth best scoring average (69.31). Chris Downes is a little-known professional who played in one tournament, the 2003 BC Open, where he finished 13-under par and tied for 18th in a weak field. The same weekend Woods played in the Open Championship, finishing fourth, with a cumulative one-over par. Woods' accomplishment that week represented better golf than Downes'—but because of differences in playing conditions directly comparing Downes' 18-under par to Woods' one-over par is not reasonable. We construct a model that accounts for the relative difficulty of each round played, and also models the whole population of golfers—placing Downes' results in perspective when compared to all players.

I use a normal distribution to model the scores for a particular player in any given round. While the normal distribution certainly is not exactly correct, it does very well at modeling scores. The distribution of scores on any one hole is difficult to model. The 18-hole round score is the sum of 18 random variables—which, based on the central limit theorem, is reasonably approximated by the normal distribution. (Tournaments are typically 72 holes in length.) In Berry, Reese, and Larkey [3], we used a normal model and found the fit to be quite good, with the residual plot showing a slightly longer right tail and a slightly shorter left tail than the normal distribution suggested.

I assume that each round has an intrinsic difficulty, labeled γ. We interpret γ to be the incremental difference from the average round on the PGA Tour. I assume that each golfer has a raw ability, θ, which is his average score in an average PGA Tour round ($\gamma = 0$). Thus, when a player with intrinsic ability θ plays a round with intrinsic difficulty γ, the mean score for that player in that round is $\theta + \gamma$. For example, if Chris Downes is playing in the BC Open, the first round difficulty may be $\gamma = -4$. If Downes has a $\theta = 73$, on average, he would shoot a 69. If he were playing The Open Championship the round difficulty might be $\gamma = 3$, which would mean that on average he would shoot 76. In other words, a 69 at the BC Open would be *equivalent* to a 76 at The Open Championship. Therefore, the mathematical model for the round score for player i, in round j, Y_{ij}, is

$$Y_{ij} \sim N\left(\theta_i + \gamma_j, \sigma^2\right),$$

where $N()$ represents the normal distribution. The parameter θ_i is player i's intrinsic mean score, and γ_j is the difficulty of round j. The standard deviation—the variation of professional golfers—is σ.

The second important aspect I need to capture is referred to as regression-to-the-mean. Let's assume all rounds are equivalent in difficulty—I still wouldn't think Chris Downes was better than Tiger Woods. Why? Because it is more likely that Downes played better than his true ability for those five rounds than that he is the best player on the planet and played average for the week. This notion is captured through a hierarchical model. In a hierarchical model a golfer's intrinsic abilities are modeled with a prior distribution. This distribution helps to statistically understand each player's performance in the context of every other player. I assume that the distribution of all the θ's on the PGA Tour is a normal distribution with a mean of μ and a standard deviation of τ,

$$\theta_i \sim N\left(\mu, \tau^2\right).$$

The standard deviation τ captures the amount of variation in true abilities for players on the PGA Tour.

A Bayesian approach to fitting the above model is selected. In this approach a prior distribution is selected for each of the parameters. Flat prior distributions are selected for each of the round parameters, γ_j, and for the standard deviations τ and σ. I utilize standard Markov chain Monte Carlo techniques to calculate the posterior distribution for each of the 1408 γ's, 2004 θ's, and the two standard deviations. The calculation techniques are not important for understanding this paper, but I do want to elaborate on one aspect of the model.

How do we estimate the difficulty of a round when we don't know the ability of the players playing the round—and conversely how do we estimate the ability of the players when we don't know the difficulty of the rounds? The model employed here is best described in an iterative fashion. First the ability of each of the players is estimated assuming all rounds are equivalent in difficulty. Using these initial estimates of player ability, initial estimates of the difficulty of each round can be made. For example, based on initial estimates of the player's abilities, if golfers averaged 2 shots higher in one round than their ability projects, then that round's difficulty is estimated to be 2. Based on the new estimates of each round's difficulty, better estimates of the player's abilities are made. This iterating of estimating the round difficulty and the player ability continues. The difficulty of each round is then directly, and simultaneously, estimated with the intrinsic abilities of the players. Simultaneously the variation of golf scores (σ) and the variation of professional players (τ) are also estimated.

The posterior mean and standard deviation of the mean score on an average PGA Tour round is reported in Table 13.1. Not surprisingly, Tiger Woods is estimated to be the best player. He is estimated to be 0.85 shots better than Vijay Singh and 0.96 better than Ernie Els. They are the only two players within one shot of Woods for an 18-hole round score. The second best player is then 3.40 shots behind Woods on average for a four-round tournament. The posterior mean standard error of an 18-hole round score is 2.81 (standard deviation of 0.005). For a four-round tournament the standard error of the total score is 5.60. The second best player is 0.61 standard errors away from Woods. There are only nine players within one standard deviation of Woods for a four-round tournament. The posterior mean for μ, the mean of the distribution of PGA Tour golfers is 73.84, and the posterior mean for the standard deviation of the distribution of golfers is 2.33. Woods is estimated to be an astonishing 5.36 shots better than the average PGA Tour golfer— in a single 18-hole round. This equates to more than 20 shots better in an 18-hole tournament.

Table 13.1. This table reports the posterior mean of the ability on an average PGA Tour round, for the ten highest rated players from 1997–2004. The standard deviation of their ability as well as the number of rounds played are also reported.

Rank	Player	θ	SD θ	Rounds
1	Tiger Woods	68.48	0.12	582
2	Vijay Singh	69.33	0.10	757
3	Ernie Els	69.44	0.13	475
4	David Love III	69.52	0.11	636
5	Phil Mickelson	69.63	0.10	635
6	Jim Furyk	69.71	0.11	695
7	Retief Goosen	69.73	0.16	272
8	Nick Price	69.82	0.12	487
9	Padraig Harrington	69.86	0.20	197
10	Sergio Garcia	69.95	0.15	331

Figure 13.1 shows a histogram for the posterior mean of each of the 2004 players mean score on an average PGA Tour round. The probability that a player of each mean skill level would beat Tiger Woods in a four-round tournament is also plotted (red line). A huge portion of the players have almost no chance to beat Woods in a tournament. It is only the very best players who have a reasonable chance to beat Woods, and those chances, individually, are small. In a typical tournament Woods has to beat 143 other golfers over four rounds. While the chance any individual player beats him is minimal, the best of 143 players would be favored. Table 13.2 shows the probability of Woods winning a tournament with different field strengths. The probability he would win a tournament in which the top 10 players, other than him, would participate, is 0.294. If the top 144 players participate, his chances decrease to 0.135. Increasing this further to a hypothetical tournament with all 2004 players, he would win with probability 0.122. Table 13.2 also reports the probability he would finish second in these tournaments. Typically this

Table 13.2. The probability Woods wins (or finishes second) tournaments with different field strengths. All 2004 represents all 2004 players from the eight-year data set derived from the tournament.

Finish	Top 10	Top 25	Top 144	$1,3,5,\ldots,287$	$1,4,7,\ldots,430$	All 2004
First	0.294	0.206	0.135	0.191	0.230	0.122
Second	0.111	0.075	0.039	0.055	0.066	0.037

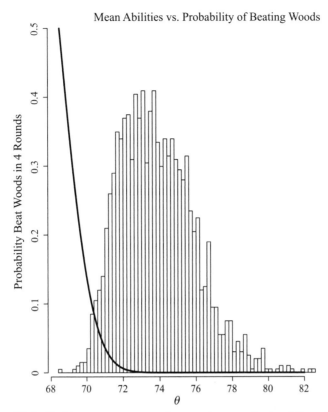

Figure 13.1. The histogram represents the posterior mean for the ability of each of the 2004 golfers in the PGA Tour data set. The line represents the probability that a golfer of the ability would beat Tiger Woods in a four-round tournament.

probability is about one-third of the chance he would win the tournament. Woods is so much better than the other players that he is the major factor in determining whether he wins the tournament. If he plays well above his ability, by random variation, he wins the tournament. This corresponds to him winning tournaments by huge margins, as he has done. When he plays poorly one of the 144 other players will beat him. If he plays a little better than average, then he may be beat by another player, and may finish second. Therefore, him finishing second is reasonably unlikely relative to winning the tournament. His 6 actual seconds is also inflated from the model prediction in part because of ties. The modeling of scores with a normal distribution assumes that there are no ties (think of it as all places playing off for the spot). In contrast, Nicklaus was a very good player, but wasn't as dominant as Woods. Therefore, his relative number of second place finishes was similar to his number of wins. As Woods ages, and invariably declines, his relative dominance will decrease, and thus his chances of finishing second will also increase.

13.3   The Career of RoboTiger

In this section we use the mathematical model for Tiger Woods, "RoboTiger," along with all of the other mathematical constructs of all 2004 players to re-play Woods' career. The model only uses the individual 18-hole scores for each round for the players. The model doesnt know how many wins each player had or what any players had done in important situations. The model is built on the scoring of each player, in each round, relative to what the competition did in those rounds. Using the model for golfers, each of the 352 tournaments are replayed 10,000 times. Of the 352 tournaments in the eight-year data set, Woods played in 147 of them, winning 36 (24.5%). In his 14-year career he played in 106 tournaments not in this data set, winning 35 of those (33.0%). In resimulating these 147 tournaments, RoboTiger won an average of 42.2 tournaments with a standard deviation of 5.4. His actual number of wins of 36 falls about one standard deviation below the average performance of RoboTiger. The true performance of Woods is quite consistent with the mathematical model of RoboTiger, if anything a little worse. Figure 13.2 shows a histogram for the number of wins for RoboTiger in the 147 tournaments. For sensitivity I ran the simulations and the modeling assuming each player had a year-specific ability. Some years, such as 2000 and 2001 Woods was more dominant than others. Assuming yearly specific θ's a resimulation of these 147 resulted in a mean of 40.3 and a standard deviation of 5.2 wins.

In addition to the 147 tournaments, I simulated each of the majors after 2004 in which Woods played (he missed two majors in 2008 with a knee injury). Since these tournaments were not in the eight-year data set, I assume that the field of these tournaments would be similar to each of the respective majors in 2004. Figure 13.3 reports the results of simulating 10,000 sets of the 50 majors for RoboTiger. The mean number of major wins for RoboTiger is 14.4 with a standard deviation of 3.2. The modal number of wins for RoboTiger is 14, which is the actual number of majors won by Woods.

Using this same approach for the majors I simulated Woods' career going forward. Assuming Woods plays to the same level for the next 10 years (he'll

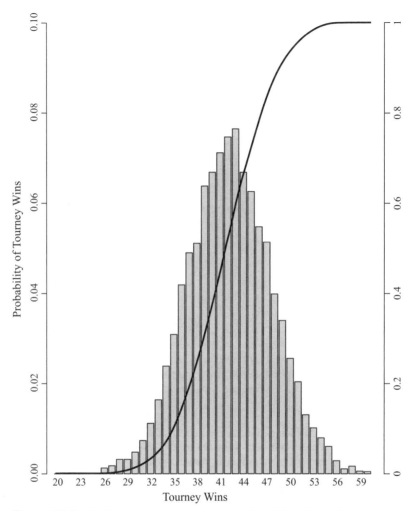

Figure 13.2. The histogram represents the proportion of times RoboTiger won each number of tournaments from the 147 played in the eight years of this dats set. Woods won 36 of these tournaments. The curve represents the cumulative probability for the number of tournament wins.

be 43) and then never wins another major, the mean number of wins is 25.5, with a standard deviation of 2.8, with an 8% chance of 30 or more, and a 99.6% chance of winning at least 19 majors, breaking Jack Nicklaus's record for all-time majors won. If he plays at the same level for 15 years, he would win an average of 29.5 majors with a standard deviation of 3.3, and a 1% chance of winning at least 19 majors. The simulations predict the 2013 US

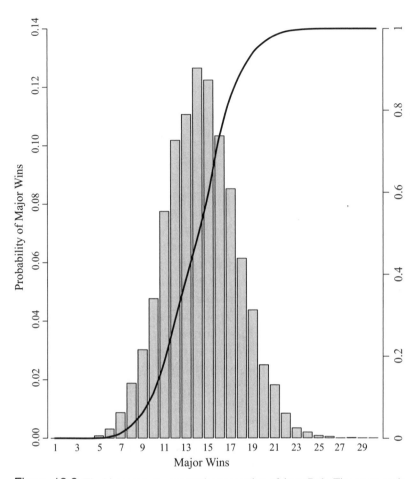

Figure 13.3. The histogram represents the proportion of times RoboTiger won each number of major tournaments in his first 50. The curve represents the cumulative probability for the number of major wins.

Open at Merion Golf Club to be the most likely major in which Woods wins the record-setting 19th major.

The number of actual career wins for Woods has been very close to the simulated career of RoboTiger. While this certainly doesn't prove that Woods doesn't have some special ability to win beyond his skill level, it does demonstrate that his record of career wins is very consistent with his skill level. The statistical modeling of Woods, and all players on the PGA Tour, has the specific characteristic that Woods does not have an innate ability to win, nor do other players have an innate decline when they are near the lead. Woods and all golfers behave as statistical robots. If this statistical modeling represents the truth for Woods, then the expectation of what would happen for his career is what we have observed. Humans have an incredible ability to create reasons for what we see. Many times there are no reasons for what we see, other than pure randomness. Many have surrounded Tiger Woods with an aura of the ultimate winner, wearing his well-known red shirt on the final day of a tournament. What we may be witnessing is not a better winner, but just a much better player than everyone else. For, if we were watching a robot with the same skills as Woods, we would have likely seen a very similar career.

References

[1] S.M. Berry, Drive for show and putt for dough, *Chance*, 12 (1999) 50–55.

[2] S.M. Berry, How ferocious is Tiger?, *Chance*, 14 (2001) 51–56.

[3] S.M. Berry, C.S. Reese, and P.L. Larkey, Bridging different eras in sports, *Journal of the American Statistical Association*, 94 (1999) 661–686.

About the Author

Scott M. Berry received his B.Sc. (Math) in 1990 from the University of Minnesota, and his M.Sc. and PhD (Statistics) in 1991 and 1994 from Carnegie Mellon University. He is President of Berry Consultants, a statistical consulting company specializing in Bayesian design and analysis of clinical trials. His interest in statistical inference grew out his love for sports and has published over 40 articles on statistics in sports. His publications vary from the *Journal of the American Statistical Association* to *ESPN The Magazine*!

CHAPTER 14

G. H. Hardy's Golfing Adventure

Roland Minton

Abstract

A little-known paper by G.H. Hardy addresses a basic question in golf: which of two golfers of equal ability has the advantage, the more consistent golfer or the more erratic golfer? Hardy models golf as a sequence of independent shots, each of which can be normal, excellent or bad. In this paper, the distributions of hole scores for simulated golfers using Hardy's rules are computed. The distributions enable us to explore Hardy's basic question in different golfing contests.

There are very few sentences in print that contain both the word "golf" and the name G.H. Hardy. Hardy (1877–1947) was one of the most prolific and influential mathematicians of the early twentieth century. His book *A Mathematician's Apology* [1] makes the case for mathematics as a pure discipline of austere beauty and uncompromising standards. He wrote, "The mathematician's patterns, like those of the painter's or the poet's, must be beautiful, the ideas, like the colours or the words, must fit together in a harmonious way. There is no permanent place in the world for ugly mathematics." He found little beauty in applied mathematics. The assumptions made by applied mathematicians, tied to the laws of physics and other mundane concerns, are not always motivated by mathematical curiosity and the results are often more pragmatic than inspirational.

Given his background as a pure mathematician par excellence, his publication in the December 1945 issue of *The Mathematical Gazette* is singular. There, on pages 226 and 227, is an article titled "A Mathematical Theorem about Golf" by G.H. Hardy [2]. He introduces a simple model of golfing and provides a preliminary analysis. Here, we discuss Hardy's results and some calculations based on his model.

14.1 Hardy's Golf Problem

The problem proposed by G.H. Hardy is motivated by a hypothetical match between two golfers of equal ability. If one golfer is much more consistent than the other, which golfer has the advantage? Golfers are often faced with a choice of attempting a risky shot or playing safely. For instance, they may choose to try to hit over a lake directly at the hole, or play safely around the lake but farther from the hole. Is it better to use a cautious strategy or a risky strategy?

Hardy approached this question by imagining a golfer whose shots are either excellent (E), normal (N) or bad (B). The golfer plays a hole with a par (target score) of four. A player hitting four normal shots (NNNN) finishes with a score of 4. A bad shot adds one to the number of shots. Therefore, a player who hits three normal shots and then a bad shot (NNNB) has not finished the hole. Another normal shot (making the shot sequence NNNBN) finishes the hole with a score of 5. By contrast, an excellent shot reduces the required number of shots by one. Thus, a shot sequence of NNE finishes the hole with a score of 3. The sequence NBEN finishes the hole with a score of 4. More examples follow.

Shot Sequence	Score
NBNNN	5
BNNE	4
NEBBBN	6

A sequence can never end in B because a bad shot always adds one to the score. However, a sequence can end in E. The sequence ENN finishes the hole with a score of 3, as does the sequence ENE. In a sense, the golfer is cheated out of the benefit of an excellent shot, because ENN and ENE receive the same score. The first two shots (EN) leave the ball close to the hole, so a normal putt from this distance will go in. An excellent putt that goes into the exact center of the hole is enjoyable to watch, but the golfer gets no extra credit for perfection.

Our most important assumptions involve the distribution of shots. We suppose that all shots are independent, so a bad shot does not affect the probability that the next shot is excellent. (All golfers wish that this was realistic.) We assume that the probability of a bad shot is p with $0 \leq p \leq \frac{1}{2}$, and the probability of an excellent shot is the same, so the probability of a normal shot is $1 - 2p$. The only difference between one such golfer (the phrase "Hardy golfer" will refer to a golfer playing with these constraints) and another is the value for p. At first glance, all Hardy golfers appear to have equal ability, since excellent shots and bad shots have equal probability.

Suppose that golfer C is a Hardy golfer with p-value p_1 and golfer R is a Hardy golfer with p-value $p_2 > p_1$. Golfer C has a higher probability of hitting a normal shot than golfer R, so golfer C is more consistent (or more cautious golfer). Golfer R has a higher probability of hitting either an excellent shot or a bad shot than does golfer C, so golfer R is more erratic (or risky). The problem is to determine which is more likely to win a match.

14.2 Hardy's Analysis

Hardy's paper presents the case where $p_1 = 0$, so that golfer C always makes a par 4. Golfer R has a probability p of hitting an E shot (which Hardy calls a *supershot*) and probability p of hitting a B shot (which he calls a *subshot*). Hardy computes the probability of golfer R winning a hole as $w(p) = 3p - 9p^2 + 10p^3$. He then computes the probability that golfer R loses the hole as $l(p) = 4p - 18p^2 + 40p^3 - 35p^4$. Details are given in [4].

The graphs of $w(p)$ and $l(p)$ in Figure 14.1 show the probabilities of winning and losing.For values of p less than approximately 0.37, player R is more likely to lose than win. The maximum vertical distance between the two curves occurs at approximately $p = 0.09$. Extensions to cases where the p-values for E and B shots are unequal are given in [3].

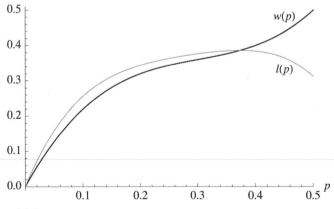

Figure 14.1. Probabilities of a Hardy golfer with parameter p winning (w) and losing (l) a hole to a par 4.

For realistic values of p, then, the consistent player is more likely to win the hole. Hardy says that this is at odds with the standard golfing wisdom that an erratic player is better off at match play (counting each hole as a separate contest) than at stroke play (where strokes are counted for all 18 holes). As we will see, the standard wisdom is actually correct, because the consistent player has an even larger advantage in stroke play.

14.3 Two Moments

To start to analyze a stroke play match between two Hardy golfers, we can calculate the mean score on a single hole. A reasonable guess is that the mean should be four, since E and B shots are equally likely. This would be correct if it were not the case that some E shots are wasted. In the sequence ENE, the second E does not improve the golfer's score, so the mean score for a Hardy golfer with $p > 0$ will be larger than 4.

We will calculate the mean and variance of the number of shots using a suggestion of Gregory Minton, by looking at the number of B shots.

If there are no B-shots, the golfer's score on the hole could be 2 (sequence EE) with probability p^2, 3 (sequences NNE, NEE or ENE) with probability $3p(1 - 2p)$ or 4 (sequence NNNN) with probability $(1 - 2p)^2$.

Generalizing, if there are k B-shots, the golfer's score could be $k + 2$ if there are two E-shots also included. One of the E's must come at the end, so there are $k + 1$ different sequences of this type. The probability of such a sequence is $(k + 1)p^{k+2}$.

The golfer's score could be $k + 3$ if there are two N-shots and an ending E. There are $\frac{1}{2}(k + 2)(k + 1)$ positions for the N-shots, so this type of sequence has probability

$$\frac{1}{2}(k + 2)(k + 1)p^{k+1}(1 - 2p)^2.$$

Further, the golfer's score could be $k + 3$ if the sequence contains one E and one N before the last shot and ends in either N or E. This has probability

$$(k + 2)(k + 1)p^{k+1}(1 - 2p)(1 - p).$$

The golfer's score could be $k + 4$ if there are three N-shots before the end, with the last shot being either N or E. The probability is

$$\frac{1}{6}(k + 3)(k + 2)(k + 1)p^k(1 - 2p)^3(1 - p).$$

These are the only possibilities. It follows that the mean is

$$\mu = \sum_{k=0}^{\infty}(k+2)(k+1)p^{k+2}$$

$$+ \sum_{k=0}^{\infty}(k+3)(k+2)(k+1)p^{k+1}(1-2p)(\tfrac{3}{2}-2p)$$

$$+ \sum_{k=0}^{\infty}(k+4)\tfrac{1}{6}(k+3)(k+2)(k+1)p^{k}(1-2p)^3(1-p)$$

and the second moment is

$$\mu_2' = \sum_{k=0}^{\infty}(k+2)^2(k+1)p^{k+2}$$

$$+ \sum_{k=0}^{\infty}(k+3)^2(k+2)(k+1)p^{k+1}(1-2p)(\tfrac{3}{2}-2p)$$

$$+ \sum_{k=0}^{\infty}(k+4)^2\tfrac{1}{6}(k+3)(k+2)(k+1)p^{k}(1-2p)^3(1-p).$$

All of the sums can be evaluated using the geometric series

$$\sum_{k=0}^{\infty}p^k = \frac{1}{1-p}$$

and differentiation. For example, multiplying by p^2 gives

$$\sum_{k=0}^{\infty}p^{k+2} = \frac{p^2}{1-p}$$

and taking two derivatives produces

$$\sum_{k=0}^{\infty}(k+2)(k+1)p^k = \frac{2}{(1-p)^3}.$$

It follows that the first sum in the calculation of μ equals

$$\frac{2p^2}{(1-p)^3}.$$

The mean and variance are

$$\mu = 4 + p\left(1 - \frac{p^4}{(1-p)^4}\right)$$

$$\sigma^2 = \mu_2' - \mu^2 = \frac{7p - 51p^2 + 156p^3 - 252p^4 + 219p^5 - 90p^6 + 10p^7}{(1-p)^8}$$

For small values of p, the mean is approximately $4 + p$ and the variance is approximately $7p + 5p^2$. The approximations are quite good for $p < 0.2$. Higher p-values are unrealistic, as $p = 0.2$ implies that only 60% of the golfer's shots are normal.

What are realistic values for p? This is an awkward question, as it implicitly grants the model with more validity than it deserves. Clearly, golf shots come in more than three categories, and especially excellent shots or egregiously bad shots do not always modify the golfer's score by exactly one stroke. Nevertheless, to get an estimate of reasonable values I computed the variance of scores for one round of some PGA tournaments. They ranged between 8 and 12, corresponding to p-values between 0.06 and 0.09.

Graphs of the theoretical distributions of scores for 18 holes with $p = 0.05$ and $p = 0.1$ are shown below. All 18 holes are par 4's played by the rules described above. The more erratic golfer with $p = 0.1$ has a wider distribution of likely scores, being more likely to shoot 69 or less and more likely to shoot 75 or higher. The mean with $p = 0.05$ is lower than the mean with $p = 0.1$. For 18 holes, they are approximately 72.9 and 73.8, respectively.

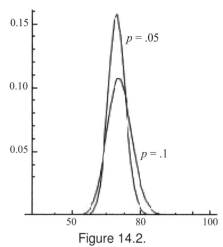

Figure 14.2.

14.4 Stroke Play

Having a lower mean score does not necessarily imply that you will win a majority of your matches. The type of match being played influences who wins. Stroke play, tournaments and skins games are discussed here. Different forms of match play are discussed in [4].

In stroke play (also called medal play), each player counts all strokes over 18 holes and the lower total wins. The probabilities of players making different scores on a given hole can be combined to compute the probability of a player having a particular score for the entire round. Comparing these probabilities, a Hardy golfer with $p = 0.05$ will defeat a Hardy golfer with $p = 0.1$ 53% of the time, with 9% ties. This is illustrated below, where the results of 20 simulated rounds are shown.

$p = 0.1$: 76,72,76,74,68,78,72,68,71,77,75,76,76,74,71,73,83,80,70,70

$p = .05$: 76,74,75,73,71,73,73,74,69,70,71,75,71,74,73,75,72,74,73,74

The more erratic golfer ($p = 0.1$) records the best score (68, twice) and the eight worst scores (76–83). If the scores are of twenty stroke play contests, the more consistent golfer has the lower score 10 times (50%) and there are two ties (10%).

In stroke play, the consistent player has a notable advantage over the erratic player, winning 53% of the matches against only 38% losses. The advantage is reduced in match play, with 46% wins against 42% losses over an 18-hole match. To find a competition in which the risk-taking erratic player has an advantage, we turn to the ultimate game for aggressive players.

14.5 Skins Game

In a skins game, all players in a foursome record their scores on a hole and the lowest score wins. If two or more players tie with the lowest score, then the group moves on to the next hole. The phrase "two tie, all tie" means that no one is eliminated if there are at least two tied for low score. There are many ways of betting on skins. The most common is to carry over all bets. If the bet is $1 per hole and the first hole is tied by two or more players, then everybody plays the second hole for $2. If the second hole is also tied, then everybody plays the third hole for $3. An erratic golfer can have a string of bad holes and still be in line to win all the money if the other three players tie every hole.

Our first analysis of Hardy golfers playing skins will look at a foursome consisting of three C-golfers ($p = 0.05$) and one R-golfer ($p = 0.1$). The calculations show that each of the C-golfers has a 8.2% chance of winning a given hole, the R-golfer has a 15.1% chance of winning it, and 60.4% of the time there is a tie. The erratic golfer has an advantage here, although with several carry-overs one of the consistent players could win all of the skins.

If the foursome consists of two C-golfers and two R-golfers, the percentages change. Each of the C-golfers has a 7.7% chance of winning a hole, each of the R-golfers has a 13.8% chance and 43% of the time there is a tie. With two erratic golfers having a chance to break loose, less than half of the skins are carried over. The number of ties depends on the composition of the foursome. The chance of a tie increases to 55% if there are three R-golfers and one C-golfer. In this case, each R-golfer has a 12.6% chance of winning a hole while the C-golfer has a 7.4% chance of winning a hole.

In all cases, erratic golfers have an advantage over consistent golfers. The actual number of carry-overs in a real skins match depends on a number of factors that we have ignored. For example, some holes are designed to tempt golfers to take large risks, effectively increasing the difference in p-values. As more holes are tied, pressure can build and affect risks that a player is willing to take. Psychologically, it is great fun to make a tying putt that keeps an opponent from winning a skin, whereas having a putt to win several skins emboldens some and tightens up others. This may increase the number of tied holes.

14.6   Tournament Golf

The other main setting in which golf is played is a tournament. The most basic tournament has everybody play one round, with the lowest score winning. Professionals typically play four rounds. Amateurs often play in tournaments in which scores are adjusted based on handicaps. Each of these cases will be considered.

Suppose we have a one round tournament contested by 70 consistent Hardy golfers with $p = 0.05$ and 70 erratic Hardy golfers with $p = 0.1$. An interesting statistical paradox is present. We have seen that the consistent golfers have a lower mean score than the erratic golfers, nearly a full stroke better on the average, 72.9 to 73.8. However, in the tournament there is a 71% chance that one of the erratic golfers will have the lowest score.

The golfers with the highest average scores are extremely likely to produce the single lowest score! The paradox is resolved by noting that the scores of the erratic golfers have a higher variance, giving the erratic golfers a higher probability of a better score. This was seen in the simulation of 20 scores each for consistent and erratic golfers. A particular erratic golfer is not a good bet to score well, but the odds are good that at least one of them will go low and have an excellent score.

The odds shift when the tournament has more rounds. An erratic golfer who gets lucky the first round is subject to the same distribution of scores for the second round, a distribution that is centered around a not-so-stellar 73.8. Over four rounds, the erratic golfers retain a slight edge, winning 60%

of the tournaments. The longer the tournament continues, the more likely it is that the steady, unspectacular golfers will prevail.

14.7 Handicaps

An important feature of the golf handicap system is revealed by the Hardy golfing model. The goal of a handicap system is to even the odds in a competition between unequal golfers. Golfer A who averages 90 could play golfer B who averages 70 if a handicap of 20 strokes were given. Then a better-than-average 87 (net 67) by golfer A would beat a worse-than-average 74 by golfer B.

The details of the handicap system administered by the United States Golf Association (USGA) are complicated (and surprisingly mathematical). Each course is evaluated and assigned a course rating and a slope rating. These determine a line $y = mx + b$ that predicts what a golfer of a given handicap would be expected to shoot on that course. A slope rating of 113 is average. On a course with course rating 70.5 and slope rating 130, we have $m = \frac{130}{113} \approx 1.15$ and $b = 70.5$, so that a golfer with handicap 10 should average $1.15(10) + 70.5 = 82$.

To compute a handicap, only the best ten of the last twenty scores (relative to the course rating and slope rating) count. The USGA's explanation acknowledges that the system tries to estimate a player's *potential* and not the typical performance.

To illustrate the difference, let us look at the simulated scores shown in the "Stroke Play" section. Assume that each round is played on a course with course rating 70 and slope rating 113. The ten best scores for each golfer are retained and then 70 is subtracted.

$p = 0.1$: 72,68,72,68,71,74,71,73,70,70 $\rightarrow$ 2, $-$ 2,2, $-$ 2,1,4,1,3,0,0

$p = .05$: 73,71,73,73,69,70,71,71,73,72 $\rightarrow$ 3,1,3,3, $-$ 1,0,1,1,3,2

The averages of the adjusted scores are 0.9 for the ($p = 0.1$) R-golfer and 1.6 for the ($p = 0.05$) C-golfer. Rounding off, the handicaps would be 1 and 2, respectively. In a match between the players, the C-golfer would get a stroke, meaning that the C-golfer's score would be reduced by 1 before comparing to the R-golfer's score. In the 20 simulated matches, the C-golfer won 10 and tied 2. Subtracting a stroke unbalances the match further, with the C-golfer winning 12 and tying 1.

This shows that in a head-to-head match the USGA handicap system favors consistent players. An erratic player is rated on the potential indicated by the ten best rounds, which can be better than average if the player's scores have a high variance. Since better players tend to be more consistent, the USGA handicap system tends to favor better players.

If you think this is unfair, consider our tournament analysis. The higher-average, erratic players dominated. Out of a large group of erratic players, it is likely that one or more will play to their potential and steal the tournament from the lower-average, more consistent players. The handicap system makes the tournament results fairer, in the sense that erratic and consistent are more equally likely to win.

14.8   Laurels to Hardy

Hardy's motivation in developing his golf model is not at all clear. This adventure in applied mathematics was uncharacteristic of his work. However, an interest in sports was not unusual. He was a devoted cricket fan, keeping up with the statistics and attending matches. So, we can imagine a question arising in conversation about the relative merits of consistency and erratic brilliance in golf. While Hardy's model is an oversimplification of golfing results, it provides insights into the role of variance in different types of golfing competitions. The insights include a clarification of the goals of the USGA handicap system, which does a better job of evening the odds in tournament play than in head-to-head competition.

References

[1] G.H. Hardy, *A Mathematician's Apology*, Cambridge University Press, 1940.

[2] ——, A mathematical theorem about golf, *The Mathematical Gazette,* **29** (1945) 226–227.

[3] G.L. Cohen, On a theorem of G.H. Hardy concerning golf, *The Mathematical Gazette*, **86** (2002) 120–124.

[4] R.B. Minton, Lipping Out and Laying Up: G.H. Hardy and J.E. Littlewood's Curious Encounters with the Mathematics of Golf, *Math Horizons*, April 2010, pp. 5–9.

[5] ——, *Golf by the Numbers*, Johns Hopkins Press, to appear.

About the Author

Roland Minton received his Ph.D. in 1982 from Clemson University, home of the 2003 NCAA golf national champions. He has taught at Roanoke College since 1986. He has co-authored with Bob Smith a series of calculus textbooks, and explored a number of fun mathematics applications ranging from sports of all kinds to chaos theory to Elvis, the calculus dog. His wife Jan also teaches at Roanoke College, his daughter Kelly teaches science in Kyle, Texas, and his son Greg is a graduate student at MIT.

CHAPTER **15**

Tigermetrics

Roland Minton

Abstract

The PGA Tour collects data on every stroke of (almost) every golf tournament, with the location of the ball determined to the inch. These data open up exciting possibilities for the analysis of golf statistics. In this paper, we present several statistics, examining various aspects of the game. Certain putting statistics indicate that professional golfers perform with greater proficiency when putting for par than when putting for birdie. A general framework for evaluating golfers at different skills is presented. Special attention is paid to Tiger Woods and his remarkable career.

Viewed one way, this article is a brief report on the most enjoyable data mining project that I can imagine. A more global, though possibly overstated, view is that this is an announcement that golf statistics are about to change dramatically. Data sets exist to do for golf what sabermetrics has done for baseball.

The data come from the PGA Tour's ShotLink system. ShotLink is a system of lasers and volunteers who record the location of every shot, including qualitative information such as lie (rough or not, uphill or not, and so on) and quantitative information such as distance to the hole (measured to the inch). We can thus answer detailed questions about the performance of (male) PGA golfers. The PGA Tour is very generous in making the data available to educators.

There is one major drawback to the data at this point. Although it includes essentially every shot taken in PGA Tour events from 2004 to 2008, the PGA Tour does not run any of the four "major" tournaments (Masters, U.S. Open, British Open, PGA Championship). The public's perception of golfers' abilities is influenced by their performances in the majors, but the majors do not contribute at all to the statistics cited here.

So, what can be learned from the data? Since approximately 1.2 million shots per year are detailed, the better question is, what do you want to know? A sample of facts follows. A more comprehensive exploration will appear in [1].

15.1 How many putts do the pros make?

In 2007, PGA golfers made 99.2% of their putts of length three feet or less. That is a high percentage, but it means that of the 168389 putts attempted from three feet or less in 2007, 1536 were missed. None of the regular tour players escaped the season without missing at least one short putt. From distances greater than three feet and less than or equal to four feet, the percentage drops to 91.5%.

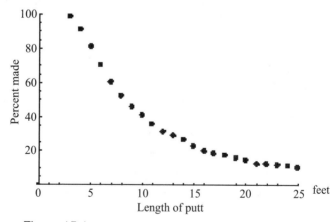

Figure 15.1. Percent of putts made from 3 to 25 feet in 2008.

Figure 15.1 shows the success rate at distances from 3 to 25 feet in 2008. The percentages in other years from 2004 to 2007 have been nearly identical. The question that most often arises is about the "break-even" point: at what distance does the percentage of putts made drop below 50%? The answer is 8 feet. That is:

> **At every distance greater than 8 feet, the pros make less than half of their putts.**

For most casual golfers, that seems like a surprisingly short distance. Of course, the pros putt on different greens than we do, they are under much more pressure, they never take mulligans and they have the PGA Tour recording every stroke, even on bad days.

15.2 Is Tiger Woods the best putter on tour?

In some ways, Tiger does not rate as a great putter. For example, in 2007 Tiger ranked 181st in percentage of putts made between 7 and 8 feet. He was 187th in putts made between 6 and 7 feet and 74th in putts made between 5 and 6 feet.

This surprises many golf fans. We are accustomed to watching Tiger make every putt down the stretch on his way to another tournament win. In particular, he seems to make every critical par putt. Maybe the word "par" in that statement is key. Does he make a higher percentage of par putts than birdie putts? The answer is "yes" from most distances in 2007, as seen in Figure 15.2.

This turned out to be a good question to ask in general. The answer is:

> **At every distance, the pros make a higher percentage of putts for par than they do for birdie.**

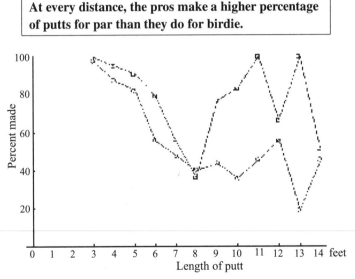

Figure 15.2. Percentage of putts made by Tiger Woods in 2007 from 3 to 14 feet, putting for par (□) or birdie (○).

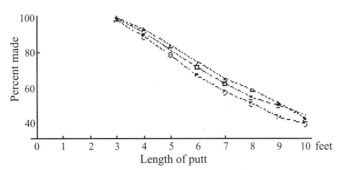

Figure 15.3. Percentage of putts made by PGA Tour in 2008 from 3 to 10 feet, putting for bogey(A), par (□), or birdie (○).

The percentage is even higher for bogey than it is for par. The tour-wide statistics for 2008 are shown in Figure 15.3. (The percentages are similar in other years.)

To return to Tiger, his best putting statistically is at the longer distances. In 2007, he ranked 8th in percentage of putts made from 10 to 15 feet and 5th in putts made from 15 to 20 feet. This does not answer the original question. Is Tiger the best putter on tour? To make progress on that, we need to answer a more important question.

15.3   What is a reasonable system for ranking putters?

There are many ways to answer this. My system starts with a compilation of tour putting averages at every distance from 3 to 100 feet. However, the averages compiled are not of the percentages of putts made. Instead, I look at the average number of putts taken when the first putt is of that distance. For example, from 35 feet the percentage of putts made is important, but so is the percentage of 3-putts. The average number of putts taken from 35 feet accounts for all results, both good and bad. In 2008, it turned out that the pros averaged 2.02 putts from 35 feet. In fact,

> **At every distance greater than 32 feet in 2008,**
> **the pros averaged more than 2 putts per hole.**

Now we can rate the putting effectiveness of a particular player. Suppose that on his first hole of the year, he starts 35 feet away and takes 2 putts. That performance is 0.02 strokes better than average. On the second hole, he starts 22 feet away and takes 2 putts. Compared to the tour average of 1.90

putts from 22 feet, that performance is 0.10 strokes worse than average. For the two holes, our pro is a net 0.08 strokes below average as a putter, or 0.04 strokes per hole worse than average. Now, do the corresponding calculations for every remaining hole that the golfer played for the year and compute that player's average performance.

In 2008, Corey Pavin's putting performance averaged 0.0547 strokes per hole or, multiplying by 18, about 0.98 strokes per round better than average. For a round, Pavin averaged almost one putt less than an average tour putter would have required putting from the same distances. Pavin's average was the best average on the tour. Tiger Woods ranked fourth at 0.84 strokes per round better than average.

By this measure, Tiger is the most consistently effective putter on tour, ranking 3rd in 2007, 17th in 2006, 8th in 2005 and 2nd in 2004. Earlier, it was noted that Tiger's "best" distances are in the 10- to 20-foot range. Since he hits a high percentage of greens in regulation, these are common putting distances for him, so his putting results from this distance are weighted heavily. The ranking system accounts for the actual distances from which each player putted during the year.

15.4 Who is the best at hitting irons from the fairway?

The ShotLink data do not include the player's club selection, so no ranking can be made of players' abilities with, for example, an 8-iron. In any case, our high-tech era of specialty clubs and hybrids minimizes the significance that can be attached to the number or symbol stamped on the bottom of the club. However, we can compile averages of how close to the hole players hit from different distances. They make it clear that Tiger is the best iron player on tour.

Since distances are measured to the inch, there is again a choice of how to split them up. I followed the PGA Tour convention of using 25-yard intervals. For example, in 2007 Tiger hit 56 shots from the fairway at distances between 100 and 125 yards from the hole. They finished a total of 11328 inches from the hole, or an average of 16.8 feet, the third best average on tour. In 2007, Tiger averaged 23.2 feet from the hole when hitting from the fairway at distances between 150 and 175 yards, the best average on tour. From distances of 175 to 200 yards, Tiger's average approach distance of 26.8 feet was also the best on tour. As the table shows, Tiger has dominated at this level for years.

Distance (yards)	2004	2005	2006	2007	2008
100-125	1	1	3	3	1
125-150	52	1	44	7	100
150-175	11	51	1	1	1
175-200	6	3	1	1	1

Figure 15.4. Tiger Woods's ranks hitting approach shots from the fairway from different distance ranges between 100 and 200 yards.

As reader Dick Green pointed out, there is a flaw in the use of average distances to rate players. Two approach shots 30 feet from the hole average the same as one approach shot 4 feet away and one 56 feet away. The second combination is likely to produce a better score. An improved rating system is given in [1].

15.5   Is there a hidden flaw in Tiger's game?

The short answer is "no!" but there *is* an interesting blip in Tiger's approach shot rankings. The distance range 125–150 yards showed some relatively large ranks in Figure 4. We can discount 2008 because of the small number of shots (19) he hit from that distance range in his injury-shortened year. (That did not stop him from dominating every other distance category.) The one year (2005) in which Tiger led the tour at the 125–150 yard distance is also the year that his 150–175 yard ranking fell significantly. So, it seems that (by his standards) there is often one distance range from which Tiger struggles.

15.6   Who is the best golfer overall?

In most sports, the ultimate product of a statistical analysis of athletes is an answer to the question of who is the best. There is little doubt in golf. Nevertheless, I have a system for rating golfers. A brief description follows. (Yes, Tiger Woods ranks number one in every year from 2004 to 2008.)

The putting rating system is described above. It gives a number of strokes better or worse than average for each golfer. The concept behind the putting system and the same measurement unit (strokes) can be extended to other types of shots.

For example, suppose that Tiger is in the fairway 122 yards from the hole, and hits his shot a mere 4 feet from the hole. The tour average from this distance (in 2008) was 22.4 feet from the hole. Both approach distances get converted to strokes. From 4 feet, the tour averaged 1.142 putts, while from 22 feet the tour averaged 1.548 putts. The conclusion is that Tiger's shot is 0.406 strokes better than average.

My system averages the player's performance (in strokes, not distance) for different types of shots. My categories include pitches (4–50 yards) from the fairway, pitches from the rough, irons (50–200 yards) from the fairway, irons from the rough, long irons (200–250 yards) from the fairway, sand shots, par-3 tee shots, par-4 tee shots and par-5 tee shots. They combine to give a grand total of strokes per round above or below average. (Details are in [1].)

For 2008, the top 5 were

Rank	Name	Rating
1.	Tiger Woods	2.65
2.	Padraig Harrington	1.25
3.	Vijay Singh	1.24
4.	Anthony Kim	1.22
5.	K.J. Choi	1.21

Figure 15.5. The top five rated PGA golfers in 2008, measured in strokes per round better than average.

In Tiger's absence, Harrington won the final two majors so his ranking as #2 is not surprising, until you remember that the rankings are based on data that exclude the majors. Harrington was obviously playing well before the majors.

Tiger's score of 2.65 means that if you add up his ratings for 2008 in the different categories, weighted by average usage (e.g., there are usually eleven par 4's on a course, so the par-4 tee shot rating is multiplied by 11), Tiger computes to being 2.65 strokes per round better than the average PGA Tour player. He also rates 1.4 strokes per round better than anyone else on tour. Over four rounds of a tournament, this predicts that Tiger wins by at least 5.6 strokes.

15.7 What else can be learned?

The possibilities are limitless. There are many questions waiting impatiently on my to-do list. For example, when a pro has a bad tournament, is it more often a result of his strengths (e.g., great putting) dropping to below average or his weaknesses (e.g., mediocre driving) becoming worse? So many questions, so little time, but we now have the data with which to find answers.

Others, including statisticians at the PGA Tour, are answering questions that would have been on my list if I had been inspired enough to think of them. This is why I suggest that there is about to be an explosion in golf statistics. The details have been recorded in the ShotLink data. All that is required is a critical mass of explorers asking good questions. I anticipate having great fun reading about all of the new discoveries in the near future.

Acknowledgements

I am indebted to Mike Vitti and Steve Evans of the PGA Tour for their time, assistance and gracious cooperation.

References

[1] Roland Minton, *Golf by the Numbers*, Johns Hopkins Press, to appear.

About the Author

Roland Minton received his Ph.D. in 1982 from Clemson University, home of the 2003 NCAA golf national champions. He has taught at Roanoke College since 1986. He has co-authored with Bob Smith a series of calculus textbooks, and explored a number of fun mathematics applications ranging from sports of all kinds to chaos theory to Elvis, the calculus dog. His wife Jan also teaches at Roanoke College, his daughter Kelly teaches science in Kyle, Texas, and his son Greg is a graduate student at MIT.

Part V

NASCAR

CHAPTER 16

Can Mathematics Make a Difference? Exploring Tire Troubles in NASCAR

Cheryll E. Crowe

Abstract

The National Association for Stock Car Auto Racing (NASCAR) has experienced tire difficulties for many years, but the problem came to the forefront at the 2008 Brickyard 400 race in Indianapolis. After a few laps, tires literally disintegrated resulting in a substantial number of laps run under caution. Because race car tires are not the same as those on personal automobiles, research and testing is needed to create more durable racing tires. The Rapid Design Exploration and Optimization (RaDEO) project has used a parameterization modification of the computer software MSC/PATRAN to explore the design of both street and racing tires. In addition, the Society of Automobile Engineers (SAE) has demonstrated the power of mathematics by creating a tire test matrix to investigate multiple variables vital to NASCAR and street tire manufacturers. This article discusses the factors involved in the design of race tires and how mathematics is used in creating and evaluating the designs.

16.1   Introduction

Trouble with tires has long plagued the National Association for Stock Car Auto Racing (NASCAR). Difficulties at various speedways across America have contributed to frustration among drivers and fans. At the 2008 Indianapolis Motor Speedway Brickyard 400 race, tires lasted only 12 laps, the equivalent of 30 miles instead of the normal 80 miles (32 laps). A record setting 52 of the 160 laps were run under caution due to the disintegration of tires.[1] Fans of the race were displeased, to say the least. With track capacity of approximately 350,000, attendance at the 2009 race was significantly low with an estimated 180,000 fans.[2] In addition, the Brickyard lost its sponsor of five years, Allstate, who announced after the race that it would not renew its contract with the Indianapolis Motor Speedway.[3]

16.2   What happened?

The problems at the Indianapolis Motor Speedway could be attributed to increasing economic difficulties throughout the United States. However, the real problem stems from a mathematical error. Goodyear, the official tire producer for NASCAR, underestimated the stress on the tires. The new heavier model car placed additional stress on the right-side tires. At the 2008 Brickyard 400 race, the tires could not withstand the pressure, and as a result, they began to fall apart after only a few miles around the track.[4]

The compound used for racing tires varies, depending on the track surface, number and tightness of turns, and banking.[5] For the Indianapolis Motor Speedway, the track is a 2.5 mile oval with 9° banking of turns, which is among the smallest banking for NASCAR tracks.[6] Depending on the tread compound, tires may provide more grip but wear faster. Additionally, the location of the tires (inside or outside) also affects wear. At the Brickyard, limited track banking, a miscalculated tire compound, and added stress to the outside tires from the new car design was a combination that led to a less than favorable outcome of the race.

16.3   Race Tires vs. Street Tires

A natural question is: what is the "big deal" about tires? If those on a personal automobile can withstand many miles and adverse conditions, what is wrong with the tires on NASCAR racers? NASCAR racing tires and the tires on personal automobiles, often referred to as street tires, are very different though the design of both is firmly grounded in mathematics.

On a typical race weekend, between nine and fourteen sets of tires are used, depending on the length of the race and type of track.[7] Tires are changed approximately every 80 miles during a race. In the 2008 Brickyard 400 race, the tires lasted only about 38% of their expected life span, resulting in at least thirteen changes per car. This is a statistically significant high number of tire changes. In comparison, the average set of street tires is replaced approximately every 50,000 miles (see Table 16.1). Additionally, the cost of NASCAR tires can be as much as twice the price of a street tire.[7] Research, development, and testing of the NASCAR tires contribute to their high cost.

Table 16.1. Comparing Racing and Street Tires [7]

	Goodyear Eagle Racing Tire	Goodyear Eagle Street Tire
Estimated Cost	$380+ each	$150–200 each
Estimated Life	80 to 150 miles	50,000 miles
Weight	24 pounds	30 pounds
Tread Thickness	1/8 inch	3/8 inch

Unlike street tires, most NASCAR teams use nitrogen in tires because it contains less moisture than compressed air. As tires heat up, they expand when moisture evaporates, causing the tire pressure to increase. Because changes in tire pressure have a significant impact on the handling of a car, teams use nitrogen for added control of tire pressure5. Street tires use compressed air as the tires do not reach the speeds of NASCAR racers and therefore evaporation is not a significant problem.

Another difference between race tires and street tires is in their composition and inflation (see Figures 16.1 and 16.2). On tracks longer than one mile, speeds are notably faster and additional tire safety precautions are required by NASCAR. An inner liner is specified, that is, essentially a second tire mounted inside the outer tire.[5] If the outer tire blows, the inner tire is still intact, allowing the driver to bring the car to a controlled stop. Air pressure in the inner liners is required to be 12–25 pounds per square inch (psi) greater than in the outer tires; for the technical inspection, left front and rear tires should be inflated to 26 psi with right front and rear inflated to 36 psi.[8] For street cars, the inflation varies based on the type and purpose of the tire.

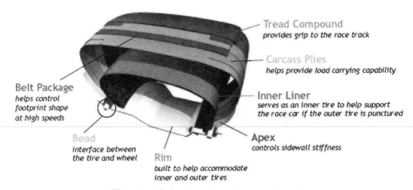

Figure 16.1. Race Tire Composition [9]

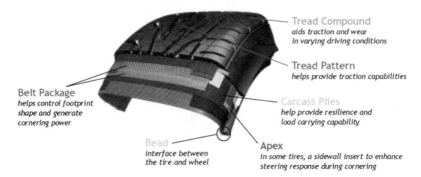

Tread Compound
*aids traction and wear
in varying driving conditions*

Tread Pattern
helps provide traction capabilities

Belt Package
*helps control footprint
shape and generate
cornering power*

Carcass Plies
*help provide resilience and
load carrying capability*

Bead
*interface between
the tire and wheel*

Apex
*in some tires, a sidewall insert to enhance
steering response during cornering*

Figure 16.2. Street Tire Composition [9]

Race tires have a flat, smooth surface called a racing slick that increases adhesion to the race track, often referred to as grip. On a dry track, tires can generate additional traction if more of the tire rubber is in contact with the ground.[5] Therefore, racing slicks provide increased contact with the track surface. Obviously, this design is inappropriate for racing during rainy conditions as a tread pattern is important in wet weather. Street tires have treads designed for different weather conditions. For additional grip, the width of racing tires is greater than that of traditional street tires (see Figure 16.3). With the increase in width of approximately four inches per tire, a racing tire provides sixteen inches of additional contact with the track and added traction in comparison to traditional street tires.

Besides the width of the tire, the depth is also significant in NASCAR. For racing tires, the smooth surface does not react in the same manner as the grooves on street tires. In particular, the tires wear differently than traditional street tires. Due to these factors, the tread compound is paramount to successful tire wear and must include a carefully constructed mixture of

Figure 16.3. Tire Surface in Race Tires and Street Tires [7]

rubber and polymer chemicals to give strength and durability while providing good grip on the track.[10] In the case of the Brickyard 400, the tire compound was inaccurately calculated resulting in the rapid disintegration of tires.

16.4   Mathematics is Making a Difference

While the majority of a car's components can be easily modeled, it can be difficult to describe tires in the same manner. Research groups are now using mathematical modeling to explore tire conditions based on multiple variables. The Rapid Design Exploration and Optimization (RaDEO) project created an automated parametric analysis to explore complex processes. In conjunction with the Ford Motor Company, Rocketdyne, and the St. Louis divisions of Boeing Aircraft Company, this project used a parameterization modification of the computer software MSC/PATRAN to explore the design of a street tire. The modification permits the use of named variables to replace the usual fixed numerical values of the modeling parameters thus allowing the mathematical models to have multiple components called functional models.[11] The Robust Design Computational System (RDCS) computer program provides for automation of design processes including parametric design scanning. Specifically, the program depends on multidisciplinary parametric mathematical models that simulate the behavior of the object being designed.

For the rubber tire model, the exploration of a 30-degree sector of the tire consisted of parametric descriptions including tread, radial ply composites, bias steel belt composite, wire bead, rim cushion, bead wrap, filler, inner liner, and chafer. In addition, the mesh and loading was parameterized to preserve sound modeling while making significant changes in the geometry of the materials as well as the overall tire size, variable inflation pressure, vehicle speed, and vehicle weight.[11] Figure 16.4 illustrates the model generated by the RDCS program.

In addition to creating a sector of the tire, the RDCS program can be adapted to create additional permutations of the base line design. The models shown in Figure 16.5 were based on the same file but included changes in the variable parameter values. The RaDEO project has demonstrated ways in which mathematics can affect the tire design process resulting in better quality street and racing tires.

The Society of Automobile Engineers (SAE) has also demonstrated the power of mathematics to affect tire design through the creation of a tire

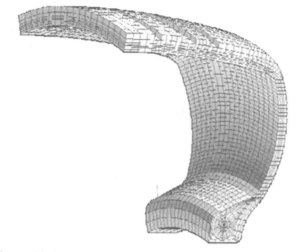

Figure 16.4. Parametric Model of 30° Sector of a Rubber Tire [11]

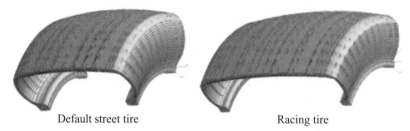

Default street tire   Racing tire

Figure 16.5. Permutations of the Base Line Design [11]

test matrix. For racing tire research, the Formula SAE Tire Test Consortium (FSAE TTC) was established to provide high quality tire data to participating FSAE teams for use in the design and setup of their racecars. The Calspan Tire Research Facility (TIRF) uses a low capacity flat-belt tire test machine that was created through the inspiration of a machine first produced by Cornell Aeronautical Laboratories in the 1960s.[12] TIRF contracts with NASCAR as well as street tire manufacturers.

Directors of the consortium, using input from the Goodyear Tire Company, developed a tire test matrix that included five components: operating conditions, relevant parameters, Calspan TIRF machine capabilities and setup, budget/time constraints, and tire popularity and availability. Parameters explored during testing included normal load, slip angle, inclination angle, slip ratio, inflation pressure, and roadway speed. Of particular interest, another "parameter" incorporated before implementing the test matrix was the warm-up rate and condition of the new tire. The initial break-in of a new tire brings the tire up to temperature as well as works the tire in such a manner that the internal molecular crosslinks and various plies rearrange into their "used" condition.[12] This is the final step in curing a tire and was completed prior to data collection.

Using the created matrix, tests were completed by the FSAE TTC to explore the interaction of parameters and to collect relevant data on the tires. Graphical outputs were produced to illustrate the interaction between variables. Of particular interest were the outcomes for lateral and longitudinal force. Three specific forces act on tires: normal force (F_z) or vertical load on the tire, longitudinal forces (F_x) or the forward and backward motion of the tire, and lateral forces (F_y) or the side-to-side movement of the tire (see Figure 16.6).

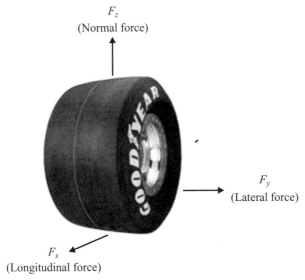

F_z
(Normal force)

F_y
(Lateral force)

F_x
(Longitudinal force)

Figure 16.6. Forces Exerted on a Tire [13]

A miscalculation of force can be linked to faulty tire design and adversely affect tire performance including grip, deformation, and disintegration. For example, the longitudinal force exerted by a tire on the wheel at the contact point is given by a characteristic function f of the tire.[14] Its components include velocity, rolling radius, and slip, which can significantly impact tire execution if plagued with mathematical errors.

Longitudinal Force: $F_x = f(\kappa', F_z)$

where:

κ is the contact patch slip $(V'_{sx}/|V_x|)$
V'_{sx} is the contact point slip velocity $(V_x r + e\Omega')$.
Wheel center longitudinal velocity in $m/s (V_x)$
Effective rolling radius (r_e)
Contact point angular velocity (Ω')
Vertical Load of the tire (F_z)

In the FSAE TTC tire tests, the relationship between lateral force and slip angle (the angular difference between the rolling direction of the tire and the plane of the wheel) as well as longitudinal force and slip ratio (locking status of the wheel) are modeled by a cubic function. The correlation between longitudinal force and slip ratio can be related to a function in the form $f(x) = ax^3$, while the relationship between lateral force and slip angle are modeled as the reflection in the form $f(x) = -ax^3$ (see Figures 16.7 and 16.8).

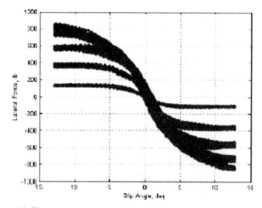

Figure 16.7. Analysis of Typical Lateral Force vs. Slip Angle [12]

Slip angle and ratio are particularly significant in racing tires. In NASCAR racers, if the turning force goes above the normal slip angle (about two to four degrees), the tires may skid too much and cause a lack of control.[15] The graphics from the FSAE TTC tire tests illustrate that a decrease in lateral force results as the slip angle increases and an increase in longitudinal force occurs as the slip ratio decreases. The longitudinal force (F_x) is approximately proportional to the vertical load (F_z) due to lateral force (F_y) generated by contact friction and the normal force (F_z).[14] The relationship is nonlinear due to tire deformation and slip.

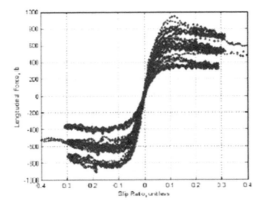

Figure 16.8. Analysis of Typical Longitudinal Force vs. Slip Ratio [12]

16.5 Problem Resolved? Looking Towards the Future

Goodyear and NASCAR believe that tire difficulties have been resolved as evidenced by the smooth running of the Brickyard 400 in 2009. The tire problems that plagued the track one year earlier were unnoticeable with a limited number of cautions and normal pit stops throughout the race. After eleven months of research and testing, it appears that Goodyear found the right compound for the racing tires.[16]

As NASCAR continues to evolve, running races under adverse weather conditions could be a possibility for the future. Currently, NASCAR suspends racing in the Sprint Cup series for rain as the tire tread and compound are not conducive to racing on a wet track. However, in other NASCAR affiliated series, races have been held during light precipitation. Most recently, part of qualifying and the final laps of the Nationwide series (considered the minor league of the Sprint Cup) were held on Sunday, August 30, 2009 during rain. While the final laps provided great excitement for spectators, some drivers are reluctant to drive in the rain when the cars are not fully equipped for it.[17] Additional research is needed to provide a safe and reasonable driving experience. The question must be asked: can mathematics make the difference?

References

[1] N. Ryan, Goodyear, Brickyard Try to Put Bad Tires in Rearview Mirror, *USA Today*. (2009), available at: www.usatoday.com/sports/motor/nascar/2009-07-23-brickyard-goodyear-tires_N.htm

[2] E. Berman, Attendance Visibly Lower at 2009 Brickyard 400. Emmis Interactive, Indiana. (2009), available at:
www.wibc.com/news/Story.aspx?id=1119359

[3] Associated Press, Allstate Not Returning as Sponsor for Brickyard 400. (2009a), available at: www.nascar.com/2009/news/business/07/28/allstate.not.returning.to.indianapolis/index.html

[4] J. Fryer, Goodyear Goofed, but NASCAR blew it on Brickyard Tire Problems, Hearst Seattle Media, LLC. (2009), available at:
www.seattlepi.com/motorsports/372571_autos29.html

[5] K. Nice, How NASCAR Race Cars Work, How Stuff Works Inc., A Discovery Company. (2009), available at: auto.howstuffworks.com/auto-racing/nascar/nascarbasics/nascar5.htm

[6] NASCARBet. NASCAR Race Track. (2009), available at:
www.nascarbet.com/nascar-race-track_1.html

[7] NASCAR.com, Race Tires vs. Street Tires, Turner Sports Interactive, Inc. (2009), available at:
www.nascar.com/news/features/tires/index.html

[8] Goodyear Racing, Fast Facts for July 24–26, 2009, The Goodyear Tire & Rubber Company. (2009), available at:
www.racegoodyear.com/news/072409.html

[9] Goodyear Global Communications, Goodyear 2009 Racing Media Kit. (2009), available at:
www.racegoodyear.com/pdf/2009GoodyearPress.pdf

[10] J. Briggs, How NASCAR Tire Technology Works. How Stuff Works Inc., A Discovery Company. (2009), available at: auto.howstuffworks.com/auto-racing/nascar/nascar-basics/nascar-tire.htm

[11] J. G. Crose, D. A. Marx, M. Kranz, P. Olson, & C. Ball, Parametric Design/Analysis with MSC/PATRAN—A New Capability, The MacNeal-Schwendler Corportation. (n.d.), available at:
www.civl.port.ac.uk/FEM/CALFEM/parametric_design.pdf

[12] E. M. Kasprzak, & D. Gentz, The Formula SAE Tire Test Consortium—Tire Testing and Data Handling, Society of Automotive Engineers, Inc. (2006), available at: www.millikenresearch.com/TTC_SAE_paper.pdf

[13] B. Volk, Tires—Cars, Trucks, You Name It, AOL. (2009), available at shopping.aol.com/articles/2009/09/17/tires/

[14] SimDriveline, Vehicle Components: Tire, The MathWorks, Inc. (2009), available at: www.mathworks.com/access/helpdesk/help/ toolbox/physmod/drive/tire.html

[15] C. Neiger, Why Doesn't NASCAR Race in the Rain? How Stuff Works Inc., A Discovery Company, (2009), available at: auto.howstuffworks.com/auto-racing/nascar/nascar-basics/nascar-rain.htm

[16] Associated Press, Jimmie Johnson Becomes Three-time Winner at Brickyard, WorldNow & WTHR. (2009b), available at: www.wthr.com/Global/story.asp?S=10791585

[17] D. Caraviello, Montreal '08 Gives Some Hope for Cup Race in Rain, Turner Interactive Sports, Inc. (2009), available at: www.nascar.com/2009/ news/headlines/cup/08/27/montreal.race.rain/ index.html?eref=/rss/news/headlines/cup

About the Author

Cheryll E. Crowe received her B.S. from Asbury College and M.A. from Georgetown College in 2003 and 2005 respectfully. She completed her Ph.D. in 2008 from the University of Kentucky with an emphasis in mathematics education. Before transitioning to Eastern Kentucky University, she taught high school mathematics for five years which sparked her interest in using technology to facilitate conceptual understanding of mathematics.

Part VI

Scheduling

CHAPTER 17

Scheduling a Tournament

Dalibor Froncek

Abstract

We present several constructions for scheduling round robin tournaments. We explain how to schedule a tournament with alternate home and away games and point out some other interesting properties of the schedules. Finally, we show what it means that two schedules are "different" and mention the number of different schedules for the cases where the number is known.

17.1 Some small tournaments

Suppose we have four teams named $1, 2, 3, 4$ and we want to schedule a three-day round robin tournament with each team playing one game on each day. All we need to do is to choose an opponent for team 1 on the first day and another opponent for team 1 on the second day. All other games are determined by these choices.

Say we choose the game $1 - 2$ for Friday and $1 - 3$ for Saturday. Of course, the remaining teams must meet on both days—teams 3 and 4 on Friday and teams 2 and 4 on Saturday. Then we have just one choice for Sunday, namely games $1 - 4$ and $2 - 3$.

Now let us try to schedule a five-day round robin tournament for six teams. More simply, we may just say "round robin tournament of six teams" as

Table 17.1. Four team tournament

	Friday	Saturday	Sunday
Game 1	$1 - 2$	$1 - 3$	$1 - 4$
Game 2	$3 - 4$	$2 - 4$	$2 - 3$

we always assume that each team plays one game each day. Because the number of opponents for each team is always one less than the total number of teams, the number of days (or rounds) is at least one less than the number of teams—we actually need to show that it is possible to find a five-day schedule. We can start as in the previous example and schedule for Round 1 games $1 - 2, 3 - 4$, and adding the game of the two new teams $5 - 6$. Then for Round 2 we schedule games $2 - 3, 4 - 5, 6 - 1$ and for Round 3 games $1 - 4, 2 - 5$, and $3 - 6$.

For Round 4 we have two choices for an opponent of team 1, either 3 or 5. Say we choose the game $1 - 3$. But then team 5 cannot play a game (see Table 17.2, since the only opponents they did not play are 1 and 3, which are scheduled to play each other! Similarly, if we choose the game $1 - 5$, we have the same problem with team 3—they could only play either 1 or 5, but these teams are already scheduled to play another game.

Table 17.2. Incomplete six team tournament

	Round 1	Round 2	Round 3	Round 4	Round 5
Game 1	$1 - 2$	$2 - 3$	$1 - 4$	$1 - 3$	
Game 2	$3 - 4$	$4 - 5$	$2 - 5$	$5 - ?$	
Game 3	$5 - 6$	$6 - 1$	$3 - 6$		

With some effort, using the method of trial and error, or, more precisely, an exhaustive (and exhausting!) search, we would find a schedule. An example is in Table 17.3.

Table 17.3. Six team tournament

	Round 1	Round 2	Round 3	Round 4	Round 5
Game 1	$1 - 6$	$2 - 6$	$3 - 6$	$4 - 6$	$5 - 6$
Game 2	$2 - 5$	$3 - 1$	$4 - 2$	$5 - 3$	$1 - 4$
Game 3	$3 - 4$	$4 - 5$	$5 - 1$	$1 - 2$	$2 - 3$

However, scheduling a round robin tournament of eight teams using an exhaustive search could take several hours. Just try to count how many possibilities we have. The first round can be selected at random. Say we select games $1 - 2, 3 - 4, 5 - 6$, and $7 - 8$.

But then just for Round 2 we have six choices for team 1 (since one out of seven possible opponents was already selected in Round 1). For Game 2 we can choose team 2, which can play one out of five opponents—if we have

Table 17.4. Games $1 - 3$ and $2 - 4$ selected in Round 2

	Round 1	Round 2	Round 3	......	......
Game 1	$1 - 2$	$1 - \{3, 4, \ldots, 8\}$			
Game 2	$3 - 4$	$2 - \{4, 5, \ldots, 8\}$			
Game 3	$5 - 6$	$6 - \{7, 8\}$			
Game 4	$7 - 8$	forced			

Table 17.5. Games $1 - 3$ and $2 - 5$ selected in Round 2

	Round 1	Round 2	Round 3	......	......
Game 1	$1 - 2$	$1 - \{3, 4, \ldots, 8\}$			
Game 2	$3 - 4$	$2 - \{4, 5, \ldots, 8\}$			
Game 3	$5 - 6$	$6 - \{4, 7, 8\}$			
Game 4	$7 - 8$	forced			

for instance scheduled Game 1 as $1 - 3$, then 2 can play any of $4, 5, 6, 7$, or 8. If we choose $2 - 4$, then 6 can select from two opponents, 7 and 8, and the last game is then left for the two remaining teams. This gives $6 \cdot 1 \cdot 2 = 12$ choices. If we choose $2 - 5$, say, then 6 can be matched with one of $4, 7$, and 8. The same would apply if we choose $2 - 7$ or $2 - 8$. This gives $6 \cdot 3 \cdot 3 = 54$ choices. Therefore, we have 66 choices just for Round 2. For each choice, there are many different choices for Rounds 3 to 6, and only Round 7 is fully determined by the previous rounds.

It is obvious that we need a better method than trial and error. To find one, we turn to the branch of mathematics called graph theory. This relatively new field (dating back to 1930s) is often used for modeling many types of applications, such as communication networks, traffic flows, task assignments, timetable scheduling, etc. Graph theory has nothing in common with graphs of functions. For us, a graph consists of a set of points, called *vertices* (singular form *vertex*), and a set of lines, called *edges*. Each edge joins two vertices. Some examples are given in Figure 17.1.

A graph in which every vertex is joined by an edge to every other vertex is called a *complete graph*. A complete graph can be viewed as a model of a round robin tournament in a natural way. Each vertex represents one team, and an edge joining two vertices, i and j, represents a game between the corresponding teams, $i - j$. If we assume that the number of teams is even, say $2n$ where n is a natural number, then a round consists of a collection of n edges such that no two edges share a vertex. Such a collection is called

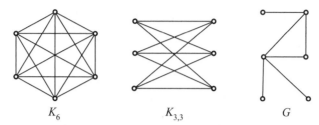

Figure 17.1. Complete graph K_6, complete bipartite graph $K_{3,3}$, graph G

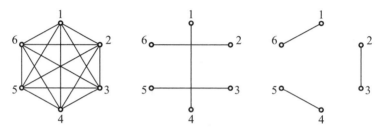

Figure 17.2. Complete graph K_6 and two different one-factors

a *one-factor* of the complete graph. An example for six teams is shown in Figure 17.2. (If two edges in the same collection share a vertex j, then team j would be scheduled to play two games in the same round, which we want to avoid.)

Now we are ready to introduce a method that is widely used in tournament scheduling. It was originally discovered in 1846 by Reverend T. P. Kirkman [6], although he did not use it for tournament scheduling, and it is probably the most popular method for tournament scheduling. When it was first used for this purpose is not known. We call a tournament using it a *Kirkman tournament*.

Construction 1 We label the vertices of the complete graph by the team numbers. Place integers 1–7 in order on a circle at a uniform distance and put the vertex 8 in the center of the circle. For Round 1 we select the edge joining 8 to 1 and all other edges perpendicular to the edge $8 - 1$. They are $2 - 7, 3 - 6$, and $4 - 5$. To select Round 2, we rotate them clockwise by $2\pi/7$. That is, we select the edge $8 - 2$ and the perpendicular edges $3 - 1, 4 - 7$, and $5 - 6$. In general, in Round k we select the edge $8 - k$ and all edges perpendicular to it.

Obviously, each team plays exactly one game in each round. To see that this construction yields a round robin tournament, we only need to convince ourselves that every team plays every other team exactly once. First we count

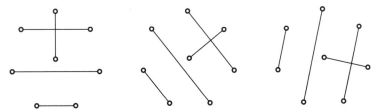

Figure 17.3. First three rounds in Construction 1

the total number of games. In each of the seven rounds there are four games, which makes a total of 28 games. This is exactly the number of games that we need to schedule—there are eight teams and each of them needs to play seven opponents. This makes 56, but we must divide it by 2 since each game was counted twice (a game $k - j$ was counted once when we counted team ks games and again when we counted team js games). We observed that every team plays a game in each round. Therefore, either every team plays every opponent exactly once, or some team misses one or more teams and hence plays another one at least twice. So we only need to show that each team plays every opponent at most once. Obviously, team 8 plays each opponent exactly once, namely team j in Round j. Now if team k plays another team, i, in round x, then the edge $k - i$ is perpendicular to the edge $8 - x$. Similarly, if team k plays i in (another) round y, then the edge $k - i$ is perpendicular to the edge $8 - y$. Because $k - i$ is perpendicular to both $8 - x$ and $8 - y$, it follows that $8 - x$ and $8 - y$ must be parallel or identical. But no two edges of the form $8 - z$ are parallel in our construction, hence $x = y$ and team k plays i in just one round, namely Round x.

Construction 1 can be easily generalized for any even number $2n$ of teams. This construction can be also described algebraically rather than geometrically. We describe two views of the algebraic construction.

Construction 2 For this construction, we define for each edge its length. Suppose we have an edge $k - j$ and both k and j are less than 8. The length of this edge is the "circular distance" between the vertices k and j, i.e., the number of steps we need to take around the circle to get from k to j using the shorter of the two paths between them. Algebraically, we define the length of the edge $k - j$ as the minimum of $|k - j|$ and $7 - |k - j|$. If we suppose, without loss of generality, that $k > j$, then the minimum is the smaller of $k - j$ and $7 - (k - j)$. We also rename team 8 to ∞ (they are pretty similar anyway) and we define the length of any edge $\infty - k$ as ∞. This definition of the edge length is consistent with the definition of edge lengths of the other edges, since $\infty - k = \infty$ for any finite k.

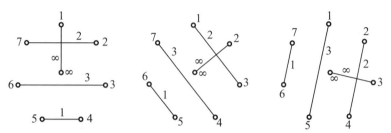

Figure 17.4. First three rounds in a Kirkman tournament

Now we observe that in Round 1 the edge $2-7$ has length $7-(7-2) = 2$, edge $3-6$ has length $6-3 = 3$, and $4-5$ has length $5-4 = 1$. Hence we have three different lengths, $1, 2$, and 3. The remaining edge $\infty - 1$ is of length ∞ and all four lengths are different. We also observe that no two vertices on the circle can be at distance more than 4, because one of the two paths between them is always shorter than four steps. Next we can construct Round 2 from Round 1 by taking each game (i.e., edge) and adding 1 to each team number except team ∞. (In fact, we can add 1 to ∞ as well, assuming naturally that $\infty + 1 = \infty$.) For doing this we need to use "wrap around arithmetic." If the finite number we receive exceeds 7, we subtract 7 to get back within our range. Game $2 - 7$ then becomes game $3 - 1$ instead of $3 - 8$ and the length of the corresponding edge is $3 - 1 = 2$, which is the same length as the length of the edge $2 - 7$. Game (edge) $3 - 6$ becomes $4 - 7$ and the length is $7 - 4 = 7$, the same as of $3 - 6$. Game $4 - 5$ becomes $5 - 6$ of length 1, same as for $4 - 5$. Finally, the edge $\infty - 1$ becomes $\infty - 2$, but the length here is defined as ∞. Hence, we again have four different lengths, $1, 2, 3, \infty$. In general, in Round $1 + i$ we add i to each vertex (team number). It is easy to observe that an edge $k - j$ of Round 1 and an edge $(k+i)-(j+i)$ of Round $1+i$ have the same length, because $(k+i)-(j+i) = k+i-j-i = k-j$.

This approach can be used to construct a different schedule. For Round

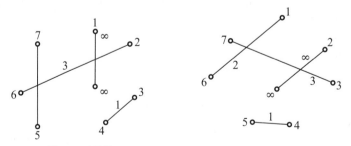

Figure 17.5. First two rounds in a Steiner tournament

1 we choose edges $\infty - 1$ of length ∞, $4 - 3$ of length 1, $7 - 5$ of length 2, and $2 - 6$ of length 3. Then for Round i we add $i - 1$ to each vertex as in Construction 1. We again have four different lengths and therefore the schedule covers each game exactly once. A tournament using this schedule is often called a *Steiner tournament*, since it can also be constructed using the so-called *Steiner triple system*.

17.2 Tournaments for any even number of teams

To make sure that this construction gives a good schedule where every team plays every other exactly once, we repeat arguments as in Construction 1. This time we discuss the case for any even number of teams, $2n$. We generalize the edge length used in Construction 2 as follows. For an edge $\infty - i$, the length is defined as ∞. For an edge $k - j$, where $j < k < \infty$, the length is the minimum of $k - j$ and $2n - 1 - (k - j)$. Obviously, the possible lengths then are $1, 2, \ldots, n - 1, \infty$.

Theorem 3 *Let K_{2n} be the complete graph on $2n$ vertices $1, 2, \ldots, 2n - 1, \infty$ and F_1 a one-factor of K_{2n} in which every edge length $1, 2, \ldots, n - 1, \infty$ appears exactly once. For $i = 2, 3, \ldots, 2n - 1$ construct a one-factor F_i from F_1 by adding $i - 1$ to all vertices $1, 2, \ldots, 2n - 1$. Then the collection $F_1, F_2, \ldots, F_{2n-1}$ contains every edge of K_{2n} precisely once.*

Proof. Each of the $2n - 1$ one-factors (rounds) contains n edges, which makes a total of $n(2n - 1)$ edges. This is exactly the number of edges in the complete graph with $2n$ vertices. To show that each edge is used exactly once, it is again enough to show that none is used more than once. First we show that, in the complete graph with $2n$ vertices, there are exactly $2n - 1$ edges of each given length l, where l is any of $1, 2, \ldots, n - 1, \infty$. For each vertex k and each fixed length l (where both $k, l < \infty$) there are two edges of length l incident with k, namely the edges $k - (k + l)$ and $k - (k - l)$, which gives $2(2n - 1)$ edges. But every edge was counted twice—the edge $k - (k + l)$ was counted once when we looked at the edges incident with vertex k and again when we counted the edges incident with vertex $(k + l)$. Therefore, there are $2n - 1$ edges of length l. For length ∞, there are also precisely $2n - 1$ edges, each one joining ∞ to one of the vertices $1, 2, \ldots, 2n - 1$.

We already have shown that the number of edges of any length is the same as the number of one-factors. We still need to show that each length is used in any one-factor exactly once. First we show that each length is used at least once. Suppose that it is not the case and length l is not used in F_i. By our assumption, there is an edge of length l in F_1, say $(k + l) - k$. Because we obtained F_i by adding $i - 1$ to all vertices of F_1, we have the edge $(k + l + i - 1) - (k + i - 1)$. But the length of this edge is also l and hence every length is present. Next we want to show that every length is used at most once. If a length l is used more than once, another length must be missing, which we have just shown cannot happen. Therefore, every length is used in each one-factor exactly once. Because the number of edges of any length is the same as the number of one-factors, it follows that each edge is used precisely once. ∎

A specific version of our theorem producing the Kirkman tournament for $2n$ teams is given here.

Corollary 4 *Let K_{2n} be the complete graph on $2n$ vertices $1, 2, \ldots, 2n - 1, \infty$ and F_1 a one-factor of K_{2n} with edges $\infty - 1$ and $k - (2n + 1 - k)$ for $k = 2, 3, \ldots, n$. For $i = 2, 3, \ldots, 2n - 1$ construct a one-factor F_i from F_1 by adding $i - 1$ to all vertices $1, 2, \ldots, 2n - 1$. Then the collection $F_1, F_2, \ldots, F_{2n-1}$ contains every edge of K_{2n} precisely once.*

Earlier we promised two different views of the algebraic construction. Here is the other one.

Construction 5 We use the same Kirkman tournament schedule as in Construction 2 and observe that in Round 1 the sum of the team numbers (not counting ∞ and its opponent) playing each other is $9 = 7 + 2 = 7 + 2 \cdot 1$, in Round 2 it is either $4 = 2 \cdot 2$ (remember in Round 2 game 2–7 becomes 3–1) or $11 = 7 + 4$, but $4 = 7 + 2 \cdot 2$. In Round 5 we have games 4–6, 7–3, and 1–2 which give the sum equal to $10 = 2 \cdot 5$ or $3 = -7 + 2 \cdot 5$. In general, in Round j the sum becomes $7\delta + 2 \cdot j$ where $\delta = -1, 0$, or 1. This is true for all Kirkman tournaments with $2n$ teams. Round j in such a tournament consists of the game $\infty - j$ and the games whose sum is $(2n - 1)\delta + 2j$ with $\delta = -1, 0$, or 1.

17.3   Some more tournament properties

If every team has its own home field, it is desirable to schedule the tournament in such a way that the home and away games for every team alternate as regularly as possible. We say that a team has a *break* in the schedule when it plays two consecutive home or away games. The most balanced schedule is one with no breaks. However, it is impossible to have a schedule in which three or more teams have a schedule with no breaks. Suppose that there are at least three such teams. Then either at least two of them start their schedules with a home game, or at least two of them start with an away game. Without loss of generality we can assume that two start with a home game. But since both of them have no breaks, they always play home at the same time and therefore can never play each other.

A Kirkman tournament has the nice property that its rounds can be re-ordered so that two teams have no breaks while all other teams have exactly one break each. We can show this using our example for eight teams. Use the convention that in a game $k-j$ the home team is j and schedule Round 1 as $\infty-1, 7-2, 6-3, 5-4$. Round 2 is then obtained from Round 1 by adding 4 to each team number (rather than 1 as in Construction 2). In general, Round $(i + 1)$ is obtained from Round i by adding 4 to each team number and the games involving team ∞ alternate ∞ as the away and home team. Then teams ∞ and 4 have no breaks, teams 1, 2, and 3 have one home break each and teams 5, 6, and 7 have one away break each. If we want to be fair and have one break for each team, we can schedule the game between ∞ and 4 in the Round 7 as $4-\infty$ rather than $\infty-4$. This also works in general for $2n$ teams, and Round $i + 1$ is then obtained from Round i by adding n to each team number.

This construction has one interesting property. When we want to schedule a tournament for an odd number of teams, $2n - 1$, we can take any schedule for $2n$ teams and pick a team j to be the *dummy team*. Whatever team is scheduled to play the dummy team in Round i then is said to have a *bye* in that round. The schedule above has the property that when we select team ∞ to be the dummy, then no team has a break in the schedule. Surprisingly, this is the only schedule for an odd number of teams with this property, as discovered by Mariusz Meszka and the author [4], who also proved that for any even number of teams there exists a unique schedule in which every team has one bye and no break.

Another important aspect of round robin tournaments is the *carry-over effect*. When there are games $k-j$ in Round i and $k-t$ in Round $(i + 1)$, we say that team t receives the carry-over effect from team j in Round $(i + 1)$. If we look at a Kirkman tournament as described in Construction 2, we can see that team 1 receives the carry-over effect from team 6 in five rounds, namely in Rounds 2 through 6, and once from team ∞ in Round 7 (there is of course no carry-over effect in Round 1). That is, team 1 plays five times during the tournament against the team that played team 6 in the previous round. This may be an advantage or disadvantage, depending on whether

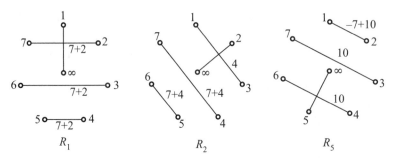

Figure 17.6. Rounds 1, 2, and 5 in Construction 5 of the Kirkman tournament

team 6 is the best or worst team in the tournament. Similarly, all teams except ∞ receive the carry-over effect from the same team four or five times,
while ∞ receives it in each round from a different team. This is due to the
rotational structure of a Kirkman tournament and the special role of ∞ in it.
There are tournaments that have *perfect carry-over effect*, that is, every team
receives the carry-over effect from each other team at most once. However,
such tournaments are known to exist only when the number of teams is either
a power of two (see [7]), 20, or 22 (see [1]). Unfortunately, the tournaments
with a perfect carry-over effect typically have very bad break structures and
balancing both properties is difficult. Some examples of leagues where both
properties are well balanced can be found in [2].

Construction 6 *(Another Schedule)* If we want to find a schedule for eight
teams different from the two schedules described in Constructions 1 and 5,
we may think of a different tournament format. We split the teams into
two divisions, East and West. First we play four rounds of interdivisional
games between teams from different divisions. After that, we play three more
rounds of intradivisional games.

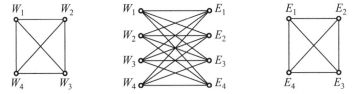

Figure 17.7. Bipartite interdivisional tournament and two intradivisional tournaments

The intradivisional part is now easy, as we already know how to schedule a tournament for four teams. A schedule for the interdivisional rounds is
not difficult either. We label the teams W_1, W_2, W_3, W_4 and E_1, E_2, E_3, E_4.

Then for Round 1 we choose the games W_1-E_1, W_2-E_2, W_3-E_3, W_4-E_4. For Round 2 we choose W_1-E_2, W_2-E_3, W_3-E_4, W_4-E_1 (note that E_4 wraps around to E_1), for Round 3 we choose W_1-E_3, W_2-E_4, W_3-E_1, W_4-E_2, and finally for Round 4 we have the remaining games W_1-E_4, W_2-E_1, W_3-E_2, W_4-E_3. This part of our schedule is called a *bipartite tournament* and the graph we use to model it is called the *complete bipartite graph* $K_{4,4}$—its vertex set consists of two disjoint sets of size four, W and E, and there are edges between every two vertices belonging the two different sets, while no edge joins any vertices within the same set.

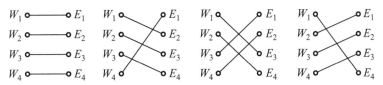

Figure 17.8. All four rounds of the bipartite tournament

One may now ask a natural question. The one-division 8-team tournament and the two-division 8-team tournament look different, but are they really? Maybe if we rename the teams in our two-divisional tournament, and re-order the rounds, we get the same tournament. If not, then the tournaments are structurally different, or, as we say in graph theory, *nonisomorphic*. In general, to distinguish whether two tournaments are isomorphic is difficult and we are not going to investigate it in full generality. However, we can convince ourselves that the two-divisional tournament is different from the one-division tournaments. It takes some time, but we can show that in the first tournament it is impossible to split the teams into two divisions such that there are four rounds with only interdivisional games and three rounds with just intradivisional games. Even Kirkman and Steiner tournaments are nonisomorphic. We can convince ourselves that any two rounds of a Kirkman tournament when put together as graphs give a cycle of length 8, while in a Steiner tournament the first two rounds give two disjoint cycles of length 4.

On the other hand, no matter how we schedule a tournament for four teams, the result is always the same. That means we can always rename the teams in one of the tournaments (mathematically speaking, renaming is what we call "finding an isomorphism") and then possibly re-order the rounds in that tournament to obtain the other one. It would take a lot of time but you can show with paper and pencil that there is only one tournament for six teams. That is, all tournaments for six teams can be obtained by re-ordering one particular tournament. Which means that you can schedule this "structurally unique" tournament in many different ways (exactly in 5!3! = 720 ways, if the order of games played on a particular day matters).

To illustrate how difficult it is to find whether two tournaments are isomorphic, we list the number of nonisomorphic tournaments for small numbers of teams: There is only one for four and six teams, there are six tournaments for eight teams, 396 tournaments for ten teams, and 526,915,620 tournaments for twelve teams. This number was found with the help of a sophisticated computer program (based on the *hill climbing algorithm*) by Jeff Dinitz, David Garnick, and Brendan McKay in 1994 [3]. In 2008, Petteri Kaski and Patric Östergård [5] used another algorithm to show that there are 1,132,835,421,602,062,347 nonisomorphic tournaments with fourteen teams! Beyond fourteen teams, the numbers are so large that only a rough estimate is known, and it is unlikely that even with the most powerful computers an exact number for sixteen teams would be determined in a near future.

There have been too many papers published about various aspects of round robin tournaments to be listed here. We refer the reader to a survey on tournament scheduling in [2].

References

[1] I. Anderson, Balancing carry-over effects in tournaments, in: *Combinatorial designs and their applications. Proceedings of the conference held at the Open University, Milton Keynes, March 19, 1997* (Fred C. Holroyd, Kathleen A. S. Quinn, Chris Rowley, and Bridget S. Webb, eds.) Chapman & Hall/CRC Res. Notes Math. 403, Chapman and Hall/CRC, Boca Raton, 1999, pp. 1–16.

[2] J. H. Dinitz, D. Froncek, E. R. Lamken, and W. R. Wallis, Scheduling a tournament, in: *The Handbook of Combinatorial Designs, 2nd edition*, (C. J. Colbourn and J. H. Dinitz, eds.), Chapman and Hall/CRC, Boca Raton, 2007, pp. 591–606.

[3] J. H. Dinitz, D. K. Garnick, and B. D. McKay, There are $526,915,620$ nonisomorphic one-factorizations of K_{12}, *Journal of Combinatorial Designs* 2 (1994) 273–285.

[4] D. Froncek and M. Meszka, Round robin tournaments with one bye and no breaks in home-away patterns are unique, in: *Multidisciplinary Scheduling: Theory and Applications* (Graham Kendall, Edmund K. Burke, Sanja Petrovic, Michel Gendreau, eds.) Springer, (2005), 331–340

[5] P. Kaski and P. R. J. Östergård, There are 1,132,835,421,602,062,347 nonisomorphic one-factorizations of K_{14}, *Journal of Combinatorial Designs* 17 (2008) 147–159.

[6] T. P. Kirkman, On a problem in combinatorics, *Cambridge and Dublin Mathematical Journal* 2 (1847) 191–204.

[7] K. G. Russell, Balancing carry-over effects in round robin tournaments, *Biometrika* 67(1) (1980) 127–131.

About the Author

Dalibor Froncek received his M.S. and Ph.D. in 1978 and 1992 from Comenius University in Bratislava, Slovakia, and his second Ph.D. in 1994 from McMaster University in Hamilton, Ontario, Canada. After that he returned for six years to Technical University of Ostrava in his home country, the Czech Republic. In 2000–2001 he was Visiting Assistant Professor at the University of Vermont. Currently he is Professor at the University of Minnesota Duluth. Besides working in graph theory and design theory, he has designed schedules for many sports leagues, including several NCAA basketball and football tournaments and some professional leagues. In 2000 he created the schedule, along with Jeff Dinitz from University of Vermont, for the inaugural (and only) season of XFL. Other professional leagues he has made the schedules for, with Mariusz Meszka from University of Science and Technology in Krakow, Poland, include Czech national soccer, hockey, and basketball leagues.

Part VII

Soccer

CHAPTER 18

Bending a Soccer Ball with Math

Tim Chartier

Aerodynamics in sports has been studied ever since Newton commented on the deviation of a tennis ball in his paper *New theory of light and colours* published in 1672. Today, the field of computational fluid dynamics (CFD) studies the effect of aerodynamics in such sports as soccer and NASCAR racing. See Figure 18.1.

Soccer matches are filled with complex aerodynamics as evidenced in the way balls curve and swerve through the air. World class soccer players such as Brazil's Roberto Carlos, Germany's Michael Ballack, and England's David Beckham exploit such behavior, especially in a free kick.

According to research by the University of Sheffield's Sports Engineering Research Group and Fluent Europe Ltd., the shape and surface of a soccer ball, as well as its initial orientation, play a fundamental role in its trajectory. CFD research has increased the understanding of the flight of a knuckleball, which is kicked so as to minimize the spin of the ball and to confuse a goalkeeper. The research group focused on shots resulting from free kicks, in which the ball is placed on the ground after a foul, for instance.

Calculating the trajectories of objects is a common problem in calculus where the absence of air resistance is generally assumed. Drag forces affect the path of a soccer ball and are of two main types: skin friction drag and pressure drag. Skin friction drag occurs when air molecules adhere to the surface of the ball, which results in friction from the interaction of the two bodies. Pressure drag occurs when the air reaches the rear of the ball. A large area then opens up for the airflow. Since the amount of moving air per unit

(a)

(b)

Figure 18.1. CFD studies aerodynamics in sports. In (a), CFD research predicts the fight of a soccer ball. In (b), a simulation of two NASCAR cars visualizes the streamlines of air produced as a car drafts and is about to pass another.

area must be constant because we are not adding or removing air the flow must slow down. Separation occurs when the air slows down so much that it is not moving or even moving backwards, which results in a wake as seen behind moving boats.

A soccer ball has a steep surface which results in a large wake; pressure drag dominates. The body of the racing car in Figure 18.1 (b) is streamlined and has less pressure drag. So, friction drag dominates.

Laminar flow occurs when streams of air flow in parallel layers. Turbulent flow is characterized by chaotic disruption between layers. Laminar flow is seen in Figure 18.1 (b) toward the front of the lead car. Turbulent flow occurs between the cars and is less visible in the picture. Both flows affect the trajectory of a soccer ball.

A turbulent boundary layer mixes air flows producing more energy close to the soccer ball. The turbulent boundary will cling to the surface longer

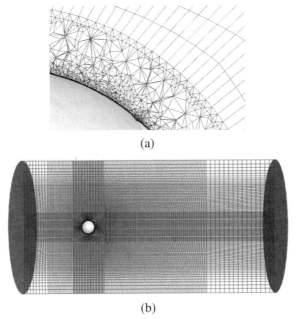

(a)

(b)

Figure 18.2. An important step in CFD simulations is capturing the geometry of a soccer ball with a 3D non-contact laser scanner. The mesh in (a) has approximately 9 million cells. As seen in (b), the space around the ball is meshed as to determine the flow of air.

and the ball will have a smaller wake. For soccer balls, a turbulent boundary layer gives a lower total drag than a laminar boundary layer. So, a transition to laminar airflow causes a soccer ball to slow down quite suddenly and potentially dip in its trajectory. The seams of a soccer ball cause more turbulence than would a perfectly smooth sphere with no seams.

With the surface so critical, an important step in CFD simulations of soccer balls is capturing the geometry of the ball with a 3D non-contact laser scanner. The mesh digitizes the surface of the ball down to its stitching as seen in Figure 18.2 (a). The refinement near the seams is required to model properly the path of air near the ball's surface. Since 1970, the official tournament ball of the World Cup has been produced by adidas for the competition, which is held every four years. From the tournament balls, four balls with different panel designs were selected to be scanned. Among the digitized balls was the adidas Teamgeist ball used in the 2006 World Cup (see Figure 18.4.

Some free kicks in soccer have an initial velocity of almost 70 mph. Wind tunnel experiments have demonstrated that air near the ball's surface changes from laminar to turbulent flow at speeds between 20 and 30 mph, depending on the ball's surface structure and texture.

Mathematically, the governing equations in such a simulation are the Navier-Stokes equations, which are based on (a) the conservation of mass, (b) the conservation of momentum and (c) the conservation of energy. The research on high-velocity, low-spin kicks did not employ (c) since an isothermal flow was assumed, meaning that the flow remained at the same temperature. The flow was also assumed to be incompressible and Newtonian. While air is a compressible fluid, this becomes influential only if a ball travels over Mach 0.3 or about 111 yards per second! A Newtonian fluid's viscosity depends only on temperature. Honey is Newtonian; if you warm it up, its viscosity decreases and it flows more easily. The viscosity of non-Newtonian fluids like shampoo or pudding depends on the force applied to it or how fast an object moves through the liquid.

These assumptions allow simplification of the Navier-Stokes equations. We again have conservation of mass in which the divergence of velocity is zero. Stated in vector form, $\nabla \cdot \overline{v} = 0$. Conservation of momentum follows:

$$\rho \left(\frac{\partial \overline{v}}{\partial t} + \overline{v} \cdot \nabla \overline{v} \right) = -\nabla p + \mu \nabla^2 \overline{v} + \overline{f}.$$

In this equation, the term on the left-hand side describes the inertia of the flow. The right-hand side of the equation is the sum of several forces. First, ∇p represents the pressure gradient, which is a physical quantity describing in which direction and at what rate the pressure changes most rapidly around a particular location. It arises from forces applied perpendicularly to the soccer ball. The second term, $\mu \nabla^2 \overline{v}$, represents the viscous shear forces in the air, which are tangential to the soccer ball. Finally, $\overline{f}$ represents other forces, usually gravity.

The techniques developed in Sheffield made possible a detailed analysis of the memorable goal scored by David Beckham of England in a match against Greece during the World Cup Qualifiers in 2001. A foul on an English player resulted in a free kick at a distance of about 29 yards from the goal. A group of defenders, the defensive wall, stood side by side on the field between the ball and Greece's goal. Beckham's shot left his foot at about 80 mph. The ball cleared the defensive wall by about one and a half feet while rising over the height of the goal. At the end of its flight, it slowed to 42 mph and dipped into the corner of the net. Calculations showed that the flow around the ball changed from turbulent to laminar flow several yards from the goal. If it had not, the ball would have missed the net and gone over the goal's crossbar.

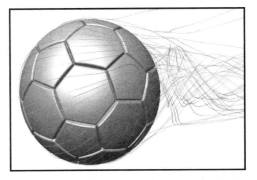

Figure 18.3. (left) Wind tunnel smoke test of a non-spinning soccer ball. (right) CFD simulation showing wake flow pathlines of a non-spinning soccer ball, air speed of 27 mph.

In a sense, Beckham's kick applied sophisticated physics. Our understanding of these dynamics could affect soccer players from beginner to professional. For instance, ball manufacturers could produce a more consistent or interesting ball that could be tailored to the needs and levels of players. Such work could also impact the training of players. Among the researchers on this project was Sarah Barber who commented, "As a soccer player, I feel this research is invaluable in order for players to be able to optimize their kicking strategies."

To this end, there is a simulation program called Soccer Sim developed at the University of Sheffield. It predicts the flight of a ball given input conditions that can be acquired from the CFD and wind tunnel tests, as well

Figure 18.4. High speed airflow pathlines colored by local velocity over the 2006 Teamgeist soccer ball.

as from high speed videos of players' kicks. The software can be used to compare the trajectory of a ball given varying initial orientations of the ball or different spins induced by the kick. Moreover, the trajectory can be compared for different soccer balls.

Analyzing aerodynamics in sports can increase the speed of a bicyclist or bobsledder or produce a more effective fastball or free kick. In CFD research, much of the work is conducted without the presence of an athlete. The impact of the CFD research of soccer balls will be seen over time and may give more insight on how to bend a soccer ball—regardless of and possibly due to its design.

Acknowledgements: Images courtesy of Fluent Inc. and the University of Sheffield. Figure 28.1(b) is courtesy of CompMechLab. The author also thanks Sarah Barber for her help on this article.

References

[1] S. Barber and T. P. Chartier, *Bending a Soccer Ball with CFD*, SIAM News **40** (July/August 2007) 6, 6.

[2] S. Barber, S. B. Chin, and M. J. Carré, *Sports ball aerodynamics: a numerical study of the erratic motion of soccer balls*, Computers and Fluids **38** (2009) 6, 1091–1100.

[3] M. J. Carré, T. Asai, T. Akatsuka, and S. J. Haake, *The curve kick of a football II: flight through the air*, Sports Engineering **5** (2002), 193–200.

[4] *Bending It Like Bernoulli*, Mathematical Moments from the AMS [online], 2008, Available:
www.ams.org/mathmoments/audioFiles/podcast-mom-soccer.mp3.

[5] *Flow Modeling Solutions for the Sports Industry from Fluent*, [online], 2006, Available:
www.fluent.com/solutions/sports/.

[6] *Sports Engineering Research Group, University of Sheffield, UK*, [online], 2006, Available: www.shef.ac.uk/mecheng/sports/.

About the Author

TIMOTHY P. CHARTIER is an Associate Professor of Mathematics at Davidson College. He is a recipient of the Henry L. Alder Award for Distinguished Teaching by a Beginning College or University Mathematics Faculty Member from the Mathematical Association of America. As a researcher, Tim has worked with both the Lawrence Livermore and Los Alamos National Laboratories on the development and analysis of computational methods to increase the efficiency and robustness of numerical simulation on the lab's supercomputers, which are among the fastest in the world. Tim's research with and beyond the labs was recognized with an Alfred P. Sloan Research Fellowship. In his time apart from academia, Tim enjoys the performing arts, mountain biking, nature walks and hikes, and spending time with his wife and two children.

Part VIII

Tennis

CHAPTER 19

Teaching Mathematics and Statistics Using Tennis

Reza Noubary

Abstract

The widespread interest in sports in our culture provides a great opportunity to catch students' attention in mathematics and statistics classes. Many students whose eyes glaze over after few minutes of algebra will happily spend hours analyzing their favorite sport. As their teachers we may use this to enhance our teaching of mathematical and statistical concepts. Fortunately, many sports lend themselves to this. This article analyzes a tennis match with a view towards its use as an aid for teaching mathematical and statistical concepts. It shows that through sports students can be exposed to the basics of mathematical modeling and statistical reasoning using material that interests them. For those who plan to become mathematics teachers, in high school or college, it provides a source of material that could enrich their teaching. It has been my experience that using sports in the classroom increases students' interest in mathematics and statistics and the public's interest in university activities.

19.1 Introduction

The difficulties faced by educators teaching mathematics and statistics are well known. To help, many textbooks try to motivate students by introducing varied applications. This addresses both students' desire to see the relevance of their studies to the outside world and also their skepticism about whether mathematics and statistics have any value. This idea works mostly with students who are committed to a particular academic or career field. For typical students, applied examples may fail to motivate if they are not of immediate concern to them or they do not occur in their daily lives.

Fortunately, students have some common interests that we can build upon when teaching mathematics and statistics. Connecting their studies to something that interests or concerns them almost always works better. Unfortunately, it is not always easy to find something that will motivate the majority of students. I have tried several things and have concluded that games and sports are the best way to accomplish this. We can help students build their studies on a foundation, an understanding of a sport that they already possess. I think this is adaptable at all levels at which mathematics and statistics is taught, from junior high to graduate schools. In what follows we discuss other advantages of using sports.

19.1.1 General

- Sports have a general appeal and scientific methods can be applied to them.

- Sports are a part of everyday life, especially for young people.

- Students usually enjoy sports and show a great deal of interest in mathematics and statistics applied to them.

- A major part of the calculus and statistics sequences offered at college level can be taught using a sport.

- Most students can relate to sports and can understand the rules and meanings of the statistics presented to them.

19.1.2 Specific

Sports data offer a unique opportunity to test methodologies offered by mathematics and statistics. I believe it is hard to find an area other than sports where one could collect reliable data with the highest precision possible. In addition to the quality measurements, here we have access to the names, faces, and life history of the participants and their coaches, trainers, and everyone involved. Almost all other data producing disciplines are susceptible to "data manipulation and data mining" and error, since unlike sports, they are not watched by millions of fans and reported on in the media. A theoretical result can be tested only when data is reliable and satisfies the conditions under which it was developed. If the validity of data produced or collected by an individual or an organization cannot be confirmed, one may end up being suspicious of the results obtained and the methodology applied.

Consider, for example, track and field. The nature and general availability of track and field data have resulted in their extensive use by researchers, teachers, and sports enthusiasts. The data are unique in that they:

1. Possess a meaning that is apparent to most people.

2. Are collected under very constant and controlled conditions, and thus are very accurate and reliable.

3. Are recorded with great precision (e.g., to the hundredth of a second in races), and thus permit very fine differentiation of change or differences.

4. Are both longitudinal (100 years for men's records) and cross-sectional (over different distances and across gender).

5. Are publicly available at no cost. Thus, they provide wonderful data sets to test mathematical and statistical models of change.

19.2 An Illustrative Example

The focus of a lesson could be on a single concept based on examples from several different sports or on several different concepts based on a single sport. In this article we illustrate how a single sport, tennis, may be used to teach mathematical and statistical concepts.

The Quirks of Scoring

During its early years, tennis used a variety of scoring systems. By the time of the first championship at Wimbledon in 1877, the All England Croquet Club had settled on a scoring system based on court tennis. This system remained unchanged until the introduction of tie-breakers in 1970.

One quirk of tennis scoring is that strange names are used for points in scoring a game: *love, fifteen, thirty, forty, game.* Although no one knows the origin of this odd system, it has been proposed that *fifteen, thirty, forty-five, sixty* were originally used to represent the four quarters of an hour. Over the years the score *forty-five* became abbreviated as *forty.* (In informal play, *fifteen* is sometimes abbreviated as *five.*) It would be simpler to score the game: *zero, one, two, three, and four.* However, the weird point names give no advantage to either player.

A more important quirk is that a game must be won by two points. If players each score three points, the score is called *deuce,* rather than *40:40.* If the server wins the next point, the score becomes *advantage in.* If the server wins again, she wins the game, otherwise the score returns to *deuce.* If server loses the next point, the score becomes *advantage out.* If the server loses again, she loses the game, otherwise the score returns to *deuce.* This feature of tennis scoring increases the chance that the stronger player will win, as we shall see.

Consider a game of tennis between two players, A and B. The progression of the game can be used to teach many statistical concepts and critical thinking. Throughout, for any event E we use $P(E)$ to denote the probability that E occurs.

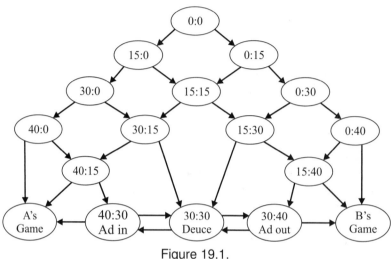

Figure 19.1.

1. Let

$$x = P(\text{A wins a point})$$
$$y = 1 - x = P(\text{B wins a point})$$

To simplify the modeling and analysis of the game we may first assume that x is fixed. Do you think this is a reasonable assumption?

Objective: **To teach critical and logical thinking versus practical significance.**

2. Starting from 0:0 what are the possible outcomes after one exchange (one point), two exchanges, and so on?

Objective: **To teach concepts such as sample space (universal set), events (sets) and their algebra.**

3. How do you assign probabilities to the outcomes of the sample space after one exchange? The possible outcomes are: 0:15 and 15:0.

Objective: **To teach concepts such as quantification of uncertainty, probability (classical, objective, and subjective), and odds.**

4. Do you think winning a point will affect the probability of winning the next point?

   Objective: **To teach concepts such as conditional probability and independence.**

5. Suppose that the answer to question 4 is no. Let $x = 0.60$. How do you find the probabilities such as $P(15 : 15)$ or $P(30 : 30)$ or $P(40 : 15)$, etc.? How do you find these probabilities if the answer to question 4 is yes?

   Objective: **To teach combinations, multiplication rule, Bernoulli, Binomial and Poisson distributions.**

6. Find

   (a) $P(\text{A wins the game without deuce})$,

   (b) $P(\text{A wins the game after reaching deuce})$,

   (c) $P(\text{A wins the game})$.

   Objective: **To teach addition rule, infinite series, and geometric progression.**

7. Find

   (a) $P(\text{A wins the set without tie-break})$,

   (b) $P(\text{A wins the set after a tie-break})$,

   (c) $P(\text{A wins the set})$.

   Objective: **To teach modeling and problem solving.**

8. Find $P(\text{A wins the match})$.

   Objective: **To teach pattern identification and model building.**

9. Find general formulas for probabilities in questions 6, 7, and 8.

   Objective: **To teach functions, graphs, and function of functions (composite functions).**

10. Let $e = x - y$ represent the edge in one point. For example, $x = 0.51$ means player A has a 0.02 edge over player B in one point. Find edges in one game, one set, and the match. Hint: call the edge in one point Δx and edge in a game Δy.

    Objective: **To teach the derivative, chain rule, and differential equations.**

11. Let $r = x/y$. Since $0 \le x$, $y \le 1$ it follows that $0 \le r < \infty$ and $r = 1$ for $x = y = 0.5$. Express the probabilities in question 9 in terms of r.

    Objective: **To teach transformations and homogeneous polynomials.**

12. Let 0, 1, 2, 3, and 4 represent the scores 0, 15, 30, 40, and the game respectively. Let

    $$g(i, j) = P(A \text{ wins the game starting from the score}(i : j)).$$

    Show that

    $$g(i, j) = xg(i + 1, j) + yg(i, j + 1)$$

    Also show that $g(3, 3) = x^2/(x^2 + y^2)$.

    Objective: **To teach recursions and difference equations.**

13. Think about a game that has reached the state deuce (or 30: 30). There is no limit to how long the game could go on. From this point, the game could reach one of the five possible states. Let 1, 2, 3, 4, and 5 denote the states: A's game, B's game, Deuce, Advantage A, and Advantage B, respectively. The game moves from state to state until one player wins. The probabilities of moving from one state to another can be summarized as

    $$\begin{array}{c} \\ 1 \\ 2 \\ 3 \\ 4 \\ 5 \end{array}\begin{array}{ccccc} 1 & 2 & 3 & 4 & 5 \\ \left[\begin{array}{ccccc} 1 & 0 & 0 & 0 & 0 \\ 0 & 1 & 0 & 0 & 0 \\ 0 & 0 & 0 & x & y \\ x & 0 & y & 0 & 0 \\ 0 & y & x & 0 & 0 \end{array}\right] \end{array}$$

    Objective: **To teach matrices, Markov chains and the states of a Markov chain.**

14. Suppose that the game is now in state deuce (state 3). This can be expressed as the state matrix;

    $$\begin{array}{ccccc} 1 & 2 & 3 & 4 & 5 \\ \left[\begin{array}{ccccc} 0 & 0 & 1 & 0 & 0 \end{array}\right] \end{array}$$

    Show that after one and two exchanges the state matrices are respectively

    $$\begin{array}{ccccc} 1 & 2 & 3 & 4 & 5 \\ \left[\begin{array}{ccccc} 0 & 0 & 0 & x & y \end{array}\right] \end{array} \qquad \begin{array}{ccccc} 1 & 2 & 3 & 4 & 5 \\ \left[\begin{array}{ccccc} x^2 & y^2 & 2xy & 0 & 0 \end{array}\right] \end{array}$$

    Objective: **To teach matrix algebra.**

15. Starting from deuce:

    (a) How many exchanges (points) are expected to be played before the game ends?

    (b) How many times is each state expected to be visited/ revisited before the game ends?

    Objective: **To teach stationary solution, inverse of a matrix, and the fundamental matrix.**

16. Suppose now that x_1, x_2 represent respectively the probabilities in part (a) and x_3, x_4 represent respectively the probabilities in part (b).

    (a) $P(A$ wins a point when serving) and $P(A$ wins a point when receiving)

    (b) $P(A$ wins a point after winning a point) and $P(A$ wins a point after losing a point)

    Find the probabilities of winning a game, a set and the match for player A.

    Objective: **To teach basic concepts of modeling.**

17. Consider a tournament like the Davis Cup. Suppose that countries A and B each have three players represented as A_1, A_2, A_3 and B_1, B_2, B_3 respectively. Suppose that the following matrix represents their chances of winning or losing against each other.

$$
\begin{array}{c c c c}
 & B_1 & B_2 & B_3 \\
\begin{array}{c} A_1 \\ A_2 \\ A_3 \end{array} &
\left[\begin{array}{c c c}
40\% & 52\% & 50\% \\
40\% & 41\% & 30\% \\
55\% & 45\% & 60\%
\end{array}\right]
\end{array}
$$

    For example, using this matrix, we have $P(A_1$ beats $B_1) = 40\%$. In the Davis Cup, each team decides which player plays the first, second, etc. game without knowing about the selection of the other team. How do you think teams should make their selection?

    Objective: **To teach game theory.**

18. In tennis, the server gets a second chance to serve after missing the first one. Ordinarily, players go for a speedy (strong) but risky first serve and a slow but a more conservative second serve. Analyze all the possible serving strategies and their consequences.

    Objective: **To teach basic concepts of decision analysis and its role in the game theory.**

19. How do you summarize statistics related to a tennis player, a team, and a tournament?

   Objective: **To teach descriptive statistics.**

20. Suppose that you have data for the speed of player A's first serve. How do you calculate the probability that in the next match the average speed of A's first serves would exceed a certain value?

   Objective: **To teach sampling distribution and central limit theorem.**

21. How do you compare two tennis players? How do you rank tennis players?

   Objective: **To teach performance measures, measures of relative standing, z-score, etc.**

22. A claim is made about the performance of a tennis player. Using the player's statistics, how do you validate the claim?

   Objective: **To teach hypothesis testing, Type I and Type II errors and P-value.**

23. How can you use the past statistics of a player to predict his or her future performance?

   Objective: **To teach estimation (prediction), confidence intervals, regression, time series, and forecasting.**

24. Suppose that you have statistics on the speed of player A's first serves. How do you predict the next record speed and perhaps the maximum possible speed of A's serves?

   Objective: **To teach theory of records, asymptotic theory of order statistics, extreme value theory, and threshold theory.**

25. How do you organize a tennis tournament?

   Objective: **To teach planning and scheduling.**

26. The winner of a men's tennis match must win three out of five sets. Each set has six games. Do you think the present scoring system is fair? For example, player A could win two sets 6-0 and lose three tie-break sets 6-7. So, A could win 30 games and lose only 21 games and yet lose the match. Do you have any suggestion to make the match more balanced?

   Objective: **To teach methods for adaptive modeling.**

19.3 Activities

We conclude with some activities that use tennis and some mathematical and statistical concepts.

Activity 1: Bouncing Ball

The balls used in different sports have a different amount of bounce. Even the balls used in the same sport may bounce differently because of their age, coverings, or simply because they contain different amounts of air. For consistency, a standard for bounciness must be established for the ball in each sport.

One way to measure the bounciness of a ball is through a quantity known as the coefficient of restitution (COR), defined as square root of the ratio of the rebound height to the initial height from which the ball was dropped:

$$COR = (H_{Rebound}/H_{Initial})^{1/2}$$

Calculation   Provide answers to the following questions.

1. A tennis ball has a COR of 0.53. If this ball is dropped from a height of 8 feet, how high will it bounce?

2. From what height should a tennis ball with a COR of 0.54 be dropped so it will bounce 12 feet?

3. Suppose that a tennis ball with a COR of 0.55 hits a wall at a speed of 65 mph. With what speed will it rebound?

Critical Thinking   Make up a question pertaining to this lesson that you, the student, would ask if you were a teacher. For example:

- You may want to know if bounciness can be quantified differently and, if so, what would be a consequence?

- Is COR independent of speed? That is, if you throw two identical balls at a wall one with the speed twice the other, would speeds of the rebounds be 2 to 1 too?

Estimation, Modeling   Select a tennis ball. Suppose that the ball is dropped from a point h feet high and has a COR equal to c.

(a) Develop a model to calculate the heights of the first, second, third, and other bounces.

(b) Think about the total distance traveled by the ball after one, two, etc. bounces. See if you could develop a model for this. Use the models you developed for a long run prediction. For example, what is the estimate for total distance traveled after n bounces?

(c) Suppose that c is unknown. How can you estimate it? For example, you may estimate COR by measuring the bounce n times and by averaging the results. Suggest a value for n.

(d) Estimate the second bounce by first directly measuring it n times and averaging as in part (c). Then apply the mathematical model in part (a) and predict the average of the second bounce. Which estimate do you prefer?

(e) Repeat part (b) for the third, fourth, and other bounces. Do you see any pattern?

Trajectories, Some Classical Functions, and Model Building Suppose now that a ball will be hit by a racket at a certain angle (similar to serves in tennis). After hitting the ground it will bounce and follow a path like a parabola (similar to the trajectory of, for example, a lob shot in tennis). Model this first for a fixed angle and fixed initial speed. Repeat the process by keeping one variable fixed and letting the other vary. Finally, let both variables vary.

Activity 2: Applying Binomial Distribution, Matrices, Markov Chain, and Derivatives

Consider a match between two players, A and B. Suppose that player A has a 10% edge over Player B in one point. That is, the probability that player A will win a point is 55% and the probability that player B will win a point is 45%.

1. Show that the probability calculations before reaching deuce can be carried out using a binomial distribution.

2. Find the probability that player A wins a game, a set, and the match given the edge A has in one point. Also calculate the edge in a game, a set, and a match given the edge in one point.

   Suppose that the information regarding the players A and B is summarized in a rectangular array as

$$
\begin{array}{ccccc}
& 1 & 2 & 3 & 4 & 5 \\
\begin{array}{c} 1 \\ 2 \\ 3 \\ 4 \\ 5 \end{array}
& \left[\begin{array}{ccccc}
1 & 0 & 0 & 0 & 0 \\
.55 & 0 & .45 & 0 & 0 \\
0 & .55 & 0 & .45 & 0 \\
0 & 0 & .55 & 0 & .45 \\
0 & 0 & 0 & 0 & 1
\end{array} \right]
\end{array}
$$

This is called a transition matrix. It includes the probabilities of moving from one state to another after a point. Here state 1 represents A won the game, state 2 represents advantage A, state 3 represents deuce, state 4 represents advantage B, and state 5 is B won the game.

3. Apply matrix algebra and interpret the results in the context of a tennis game.

4. If we look at the tennis game as a Markov chain, what are the states? Which states are non- recurrent? Which states are recurrent? Which states are absorbing?

5. Let x denote the probability that player A wins a point, and $y = 1 - x$ denote the probability that player B wins that point. It can be shown that

$$P(\text{A wins the game}) = x^4\left[1 + 4y + 10y^2 + 20xy^3/(x^2 + y^2)\right].$$

Replace y by $1 - x$ in this equation and find its derivative with respect to x. Calculate the value of the derivative at the point $x = 0.50$. For players close in ability (when the edge in one point is small, e.g., 1%) the resulting value provides the edge in one game. Compare the value obtained using the derivative with the actual value of the edge.

6. Consider the equation in problem 5. Replace y by $1 - x$. The resulting function has several properties. For example, the function is symmetric with respect to $x = 0.5$. Study the other properties of this function.

7. Consider the formula in problem 5. Replace y by $1 - x$. Suppose that $P(\text{A wins the game}) = 0.60$. Use numerical methods to find x.

8. Find the probability of winning a set as a function of x and show that it is an example of a function of a function. Use it to find the edge in a set both directly and by using the derivative (chain rule) as in problem 5.

Activity 3: Calculations Based on Normal Distribution

1. Suppose that the average speed of a tennis player's first serve is 117 mph with a standard deviation of 5 mph. What is the probability that this player's next first serve will be:

   (a) Slower than 115?
   (b) Faster than 120?
   (c) Between 116 and 122?

2. Suppose that tennis balls are produced to have COR $= 55.5\%$ (target value). To see if the process is on target once a day 50 balls are tested. If the average COR falls outside the interval 53.5% and 57.5% the process is judged out of control. What is the probability that the process will be judged out of control incorrectly? Assume that the standard deviation is 1.5%.

Activity 4: Constructing Confidence Intervals and Testing Hypotheses

1. Look up statistics for the number of aces made in 36 matches by a tennis player of your choice. Construct a 95% confidence interval for the average number of aces for this player. Hint: Use the Central Limit Theorem.

2. A sample of 36 serves of a top player on hard court has mean of 107 mph and standard deviation of 6.5 mph. His coach claims that the average speed of his serves is 110 mph. Check to see if data supports this claim. Use a 0.05 level of significance.

3. Another sample of 36 serves from the same player (problem 2) on a grass court has mean of 105 mph and standard deviation of 10 mph. Can we conclude that the mean speed of his serves on the hard court is greater (at the 0.05 level) than the mean speed of his serves on the grass court?

4. Construct a confidence interval for the difference between the population means in problems 2 and 3.

Activity 5: Applying Regression and Time Series for Prediction

1. Use statistics for a player of your choice who participated in the latest Wimbledon tournament. Use regression to predict the total points

won using, for example, the number of opponents' unforced errors as predictor. Try other factors to see if you can find the best predictors.

2. If we cannot apply regression, we can still use smoothing techniques for prediction when the data form a time series. Time series refers to data with a time index. Use smoothing to predict the number of matches a player of your choice may win next year using the number of matches the player has won in previous years.

Activity 6: Research topics

The analysis of a tennis game can be expanded in many different directions. Examples include Canadian doubles or cut-throat, Australian doubles, regular doubles, and even a five-way game when server sits out for a game. Examples of research topics include determination of the size of a handicap to make a game a fair game and analyzing of the methods used for ranking tennis players.

Acknowledgments I would like to thank Professor Joe Gallian for providing me with help and direction.

About the Author

Reza D. Noubary received his B.Sc. and M.Sc. in Mathematics from Tehran University, and M.Sc. and Ph.D. in statistics from Manchester University. His research interests include time series analysis, modeling and risk analysis of natural disasters, and applications of mathematics and statistics in sports. He is a fellow of the Alexander von Humboldt. He frequently teaches courses based on sports. His outside interests include soccer, racquetball, and tennis.

CHAPTER 20

Percentage Play in Tennis

G. Edgar Parker

Abstract

Tennis scoring makes tennis different from most games in that scoring at the point level is not cumulative and it is thus possible for a player scoring fewer points than her opponent to win a match. In this paper, a probabilistic model for a tennis match is constructed in which the probabilities associated with winning points are used to analyze the chances of winning matches. Charts are used to illustrate the consequences of the model and some numerical outcomes are highlighted to indicate how raw match data might be used to indicate how a player might improve his game. The notion of a "big" shot is examined within the model and the possibility of a "big" shot turning a losing game into a winning game is addressed.

20.1   Introduction

One of the intriguing supplements to television commentary for tennis matches these days is the computer analysis from software packages that compile the results from the charting of matches. Charting is not new, but the facility that the computer provides in creating, tabulating, and cross-referencing data from the charts has made more information easily accessible. The computer can use the charts to provide ready information on the progress of a match or data for the analysis of a player's strengths and weaknesses over several or many matches. Furthermore, software that can provide such data is readily available. A fundamental question that follows is "What does a player need to know to put the data to best use?"

If the commentators who do tennis on television or the writers who do analysis in the tennis magazines are representative, some grave misunderstandings exist as to what the mathematical consequences of the percentages associated with stroke production are. The big question may not be "How well does a player play the big points?"; it is probably "How well does a player have to play to get to play a big point?". (Vic Braden would probably add "more than once a summer?"!) In *Tennis for the Future*, Braden emphasizes that the laws of physics are a given; a player can either fight them or use them to his or her benefit. The laws of probability are also a given. A better understanding of tennis from a mathematical point of view can give a player a clearer picture of how good she needs to be relative to the strength of the opponent in order to get to the point that strategy has a reasonable chance of making a difference.

A mathematics/tennis intuition test:

1. Suppose that a player expects to win, on the average, one point for each three the opponent gets on serve. How often during the course of a match should the player expect to be able to break serve?

   A. one game in three?

   B. one game in six?

   C. one game in twenty?

2. Suppose that a player expects to win four points out of five on serve. How frequently can the player expect to win six consecutive service games?

   A. four times out of five?

   B. two times out of three?

   C. three times out of five?

If you took the test and answered "one game in twenty" to the first question and "none of these, they're all too high" to the second; your tennis intuition is in line with the mathematical realities. If you missed one of the questions, however, you can probably benefit from learning how to make a mathematical model of a tennis match. After all, a player's view of how a tennis match will go is likely biased by how the player would like it to go. What's more, understanding what a model predicts can put a player or coach in a position to answer questions that practice can help.

20.2 The Model

The basic element of tennis scoring is the point, but tennis matches are won by the player who wins two out of three (or three out of five) sets — a scoring tally three steps removed from the point. Players can practice making shots which can win points. In order, however, to plan on winning matches, a player must understand the cumulative impact that winning points is likely to have on winning matches. The initial assumption for our model is that there exists a source for knowledge of the expectation of winning any one point. This is a good assumption mathematically because it begins with the basic element of scoring and, from it, conclusions about the likely outcomes of games, and therefore sets and matches are available. It is a good assumption practically because data can be gathered whenever players have played each other, and these data can be used to suggest plausible probabilities. What's more, data can be collected that will afford analysis of parts of the game which can be directed at individual strokes or special tactics.

Suppose that a model is to be made to prepare for an upcoming match between two players. To begin, assign to each player the probability of winning a point against the other; that is, represent the probability of the first player winning a point against the second and vice versa, with non-negative numbers that sum to 1. The second assumption for the model is that a player is as likely to win any one point in a game as any other. Mathematically this makes the model computationally tractable. Practically it eliminates factors such as momentum, changes in circumstance, or "mental toughness." The assumption allows the model to act as if there are no corrupting intangibles involved and that the match will be played under similar conditions as before, and thus to predict what is likely to happen based on previous experience. (Ironically, eliminating intangibles from the model can provide a baseline against which the existence of intangibles such as "playing the big points well" can be studied scientifically.)

20.3 The Calculations

The model begins with two probabilities that sum to 1 and the assumption that the probability for winning a point does not depend on the outcome of the previous point. Represent these two probabilities by u and f, u for the likelihood that the underdog will win a point and f for the likelihood that the favorite will win a point. In traditional scoring, there are four ways to win a game: at love(0), 15, 30, or after deuce. Thus the probability of the underdog winning a game can be calculated by computing the separate probabilities of winning each type of game and then summing these probabilities. For a love game, one player must win all four points, so for the underdog this gives a probability of u^4. For a game at 15, the probability of the underdog winning is $4u^4 f$ (win the fifth point and three of the other four; that is, win the first three and the fifth points, or win the first two and last two points, or win the first point and last three points, or win the last four points). For the underdog to win a game at 30, the player must win the sixth point and three of the first five for a total probability of $10u^4 f^2$. The probability of the underdog winning a deuce game is $(20u^5 f^3)/(1 - 2uf)$. In Van Alen (no-ad) scoring, if the score is tied at three points for each player, instead of playing until one player achieves a two-point advantage, a single point is played, with the receiver choosing court, to decide the winner of the game. If Van Alen scoring is used, the probability of the underdog winning a no-ad game replaces that of a deuce game and is $20u^4 f^3$. Thus the probability of winning a game can be computed given the probability of winning a point. Table 20.1 shows comparative values that come from the formulas.

Entries in the table may be interpreted relative to some match situations. What might give the illusion of an underdog giving someone a good match? How about winning every other point, that is, getting to 30-40 consistently? This means winning two points for each three the opponent wins, corresponding to a probability of .4. This forecasts winning about 1/4 of the

Table 20.1.

Probability underdog will win point	Probability underdog will win game (standard scoring)	Probability underdog will win game (Van Alen scoring)
.05 (1 point in 20)	.00009232	.000193
.1 (1 point in 10)	.0014478	.002728
.15 (3 points in 20)	.0071371	.012103
.2 (1 point in 5)	.0217788	.033344
.25 (1 point in 4)	.050781	.070557
.3 (3 points in 10)	.099211	.126036
.33 (1 point in 3)	.139081	.168214
.35 (7 points in 20)	.1703553	.1999846
.375 (3 points in 8)	.21462	.243021
.4 (4 points in 10)	.264271	.289792
.45 (9 points in 20)	.37685	.391712
.46	.4	.413
.47	.4254	.4346
.48	.45	.456
.49	.475	.478
.5	.5	.5

Table 20.2.

Probability of winning point	Probability of winning set	Probability of winning match (2 out of 3)	Probability of winning match (3 out of 5)
.05	1.3×10^{-22}	tiny	really tiny
.1	1.928×10^{-21}	1.1×10^{-29}	tiny
.15	2.73×10^{-11}	2.246×10^{-21}	2.05×10^{-31}
.2	.000000021534	1.39×10^{-15}	10^{-22}
.25	.0000033	3.2646×10^{-11}	3.6×10^{-16}
.3	.0001161363	.0000000843	4.71×10^{-11}
.33	.001161363	.00000404316	.0000000156
.35	.003605583	.00003890694	.0000004662
.375	.01255762	.0004691209	.00001943146
.4	.03652245	.003904234	.00046087
.45	.1846393	.08968572	.04680059

games, or about four out of sixteen. That's losing 6-2, 6-2! How about the probability associated with losing deuce games on the first deuce? (Consistently getting to deuce would justify the illusion of being competitive.) Losing a one-deuce game means winning three points while losing five and corresponds to a probability of .375. In this case, on the average, the underdog can expect to win about one game out of five; that's a 6-1 set for each 6-2 set! There is a lesson here : Being competitive in individual games does not translate into any realistic hope of winning sets, much less matches. (Nevertheless, comparing the far right-hand column to the middle column, given the choice, an underdog should prefer Van Alen scoring to standard scoring.) To re-emphasize the consequences of the data in Table 20.1, Table 20.2 contains data that relate probabilities of winning a point to probabilities of winning a set or a match.

These numbers tell a sad story. Unless a player can win almost half the points, the player is going to lose the match. To get to the point of occasional upset (just under one two-out-of-three match out of every twelve played), a player needs to expect to win nine points for every eleven of the opponent.

20.4 Big Shot Strategies

Fortunately for those who might not typically go on court expecting to win more points than the day's opponent, there is relief from the numbers. The model employed so far does not take into account the conditional probabilities associated with the "big" shot, a shot that, if a player gets to hit it, increases the chances for winning the point considerably. To make a model that will assess the impact of such a shot, three qualities need to be known: the "effectiveness" the shot is likely to have, the likelihood the player will get to hit the shot, and the likelihood the player will win the point if he doesn't get to hit the shot. All of these factors can be quantified from match charts. A reasonable way to model effectiveness is to count points in which the stroke being studied is hit and note the outcomes. An effectiveness probability can then be computed by dividing the number of points during which the shot was hit into the number of points won during which the shot was hit. To model the likelihood that a shot will be hit, divide the number of points observed into the number of points in which the shot being studied was hit. Finally, to model the effectiveness (or lack of effectiveness) when the shot is not hit, divide the total number of points in which the shot being studied was not hit into the number of points won in which the shot was not hit. (This idea can be expanded to include combinations of shots.)

Let A represent the probability that the shot being studied will be hit, E represent the effectiveness probability of the shot, and E' the effectiveness probability when the shot is not hit. The probability of winning a point is $(AE) + \big((1 - A)E'\big)$. Since, from previous considerations, it is clear that a player must win more than half of the points to win consistently, the focus belongs on the relationships among A, E, and E' in the formula and a search for values for which the formula yields a value of at least .5. Assuming that $E > E'$ makes sense[1] because if the shot being analyzed is not more efficient than the player's game without it, then it hardly qualifies as a "big" shot. Mathematically, this has the useful side effect of guaranteeing that increased opportunity or efficiency means increased expectation of winning points. This enables computation of values for A, E, and E' so that raising the value of A or E while maintaining the value of E' results in a higher expectation of winning a point. Hence the basis for the analysis can be focused on computing values for A, E, and E' where the expectation of winning a point is exactly .5 with the assurance that any higher values of A and E will increase the expectation of winning. Table 20.3 lists values for A, E, and E' in which $E > E'$ and the resulting probability of winning a point.

The values in the third column were chosen to conform to the game situations discussed earlier. For .375 and .4, the opportunity levels associated

---

[1]The assumption that $E \leq E'$ may be pertinent for other investigations however. In particular, assessing the relative payoff in improving a stroke which is currently a liability would fall under this assumption.

Table 20.3.

Expectation of hitting shot	Effectiveness with shot	Effectiveness without shot	Probability of winning point
.2	1	.375	.5
.39	.7	.375	.5
.455	.65	.375	.5
.17	1	.4	.5
.33	.7	.4	.5
.4	.65	.4	.5
.09	1	.45	.5
.2	.7	.45	.5
.25	.65	.45	.5
.02	1	.49	.5
.05	.7	.49	.5
.06	.65	.49	.5
.15	.55	.49	.5

with the efficiencies that can turn the game in favor of the big shot are pro-hibitively high. On the other hand, at the level starting with .45, an absolute putaway shot (effectiveness 1) need occur in less than one point out of ten in order to turn the probabilities. In addition, shots lethal enough to participate in winning two points out of three need be hit only one point in four. At the levels of real competitiveness, even a small change in effectiveness is asso-ciated with a realistic opportunity level. The numbers again speak clearly: A game that is competitive at the point level, but not at the match level, can be made competitive with a "big" shot.[2] A game that is competitive at the game level can be made competitive with only marginal improvement of some shot.

The challenge here tactically is to determine which shots in a player's repertoire are "big" shots, what the player's efficiency level is without them, and how to optimize the opportunity to hit them. Opportunity is important. Suppose, for instance, that a player has a great overhead. Unless the player's opponent is willing to cooperate and hit some lobs no one will ever find out. Suppose that a player has a great cross-court forehand from mid-court. Unless the player's opponent is willing to hit a short ball on the correct side of the court no one will ever know. If a player's "big" shot is a groundstroke,

---

[2] It is ironic that playing the percentages has become synonymous with keeping the ball in play.

the player had better learn patience and hit deep ground strokes so that she will be able to wait for an opportunity to hit the "big" shot. On the other hand, if a player's "big" shot is a volley, the player gets to control the opportunity to hit the shot since he controls when to come to net.

Then, of course, there is the serve, the one shot whose opportunity to play is completely under a player's control and that affords a natural approach shot behind which to volley. A player gets to, within the rules, choose where to stand and how to toss the ball; the player's opponent cannot dictate what stroke the player will use nor to which part of the service box the ball will be hit. The server exercises complete control over one of the three factors in the formula, namely how often she gets the opportunity to play the chosen stroke. The model can be easily refined so as to account for double faults by using the percentage of second serves in, A', with the factors defined before to give the expected probability of winning a point as

$$(AE) + \big((1 - A)A'E'\big).$$

In this context, A, the likelihood the server will get to hit the shot, is the percentage of first serves in and E is the probability of winning the point if the server's first serve goes in.

To make sure that the data this formula can give is applied plausibly, the analysis must be different than that made earlier. Whereas before any point in the set was modeled, and thus showed what the set as a whole looked like, now only points on serve are modeled and a player gets to serve only every other game. One way to account for this is to analyze the likelihoods associated with playing six games, since by the time six service games are played, the set is over or it is time to play a tiebreaker. If a player can make some assessment as to her ability to break serve, such information can be used to assess the overall chances of winning.

Table 20.4 is made by starting with the probability of winning a game on serve. The pertinent formulas are, with p representing the probability of winning a game, p^6 for winning six in a row, $6(1-p)p^5$ for winning exactly five of six, and $15(1-p)^2 p^4$ for winning exactly four of six.

Table 20.4.

Probability of winning game	Probability of winning 6 of 6	Probability of winning 5 of 6	Probability of winning 4 of 6
.55	.02768	.16	.44
.58	.038	.2	.5
.74	.164	.51	.81
.89	.497	.86553	.98

The analysis creates a position from which to study serve as a "big" shot. Game probabilities of more than .58 mean a player should expect to win four of six more than half the time, more than .744 means a player should expect to win five of six more than half the time, and more than .89 means a player should expect to win six of six more than half the time. Going back to the work done in linking point expectation to game expectation, the critical point probabilities are, respectively, .535 for winning 58% of the games, .605 for winning 74% of the games, and .7 for winning 89% of the games. Table 20.5 gives service data and the associated point probabilities that come from the model. The chart is divided into four parts. The first treats in-percentages and efficiency levels at the break-even mark. The other three give point probabilities associated with the game probabilities from the previous tables.

Table 20.5.

Probability 1st serve in (A)	Effectiveness of 1st serve (E)	Probability 2nd serve in (A')	Effectiveness of 2nd serve (E')	Probability point won
.35	.95	.7	.4	.514
.4	.85	.7	.4	.508
.45	.8	.7	.4	.514
.5	.75	.7	.4	.515
.55	.7	.7	.4	.511
.25	.95	.7	.5	.5
.3	.85	.7	.5	.5
.35	.8	.7	.5	.5705
.4	.75	.7	.5	.51
.45	.7	.7	.5	.5075
.5	.65	.7	.5	.5
.3	.95	.8	.4	.509
.35	.85	.8	.4	.505
.4	.8	.8	.4	.512
.45	.75	.8	.4	.513
.5	.7	.8	.4	.51
.2	.9	.8	.5	.5
.25	.8	.8	.5	.5
.3	.75	.8	.5	.505
.35	.7	.8	.5	.505
.4	.65	.8	.5	.5

Before completing the chart, note what the numbers say about the first serve as an equalizing "big" shot. Suppose that point expectation is examined by assuming that a player hits his second serve stroke as a first serve as well. A second serve that goes in 70% of the time and is 40% efficient (.7 and .4 in the chart) gives a point probability of .364, for .7 and .5, a probability of .455, for .8 and .4, a probability of .384, and for .8 and .5, a probability of .48. All of these are losing games. Even the most miserable of these, however (.7 and .4) can be turned into a winning game if the player has a first serve that she can get in two times out of five and win behind 85% of the time it goes in. The numbers also indicate that double faults are not nearly as disastrous as they are often considered to be. Compare the probability associated with .7 and .5 with that from .8 and .4 by taking one more double fault for each ten second serves and trade it for one more point out of each ten second serves in. .455 is much greater than .384!

The data from the second part of the table give a baseline from which to evaluate a player's service. The data given is associated with high efficiency levels since the premise is that serve is a "big" shot. For second serve levels that are close to being winning games (.7, .5; .8, .5; and .9, .5) the first serve percentages associated with winning five of six half the time or six of six half the time are attainable. The numbers speak clearly once again. From the first part of the chart: A player can win games with a strong first serve. But from the second part of the chart: If a player wants to win sets, his second serve should break even by itself and first serve must win points when it goes in.[3]

20.5 Analyzing Serve

In practically any televised match in which a big server begins to struggle, the commentator in charge of professional expertise is almost certain to say something on the order of "Isner should take something off and just get his first serve in." The service model we have developed enables an appraisal of whether this is a good idea in the long run.

The likelihood of winning a point on serve is $(AE) + ((1 - A)A'E')$. E is larger than E', or else the second serve stroke is more effective than the first serve stroke and anyone should realize that he should be hitting it all the time. Similarly, A' should be larger than A, since if a player could get

---

[3] Readers who are old enough may remember when Ivan Lendl made the jump from Top-10 player to #1 (where he remained for some time). Commentators tended to suggest that it had something to do with mental toughness. I suggest, in light of the analysis above, that the fact that he improved his first serve enough to give him some aces and maintained his ground game at the level he had previously played it pushed him to the top.

Table 20.5. (continued)

Probability 1st serve in (A)	Effectiveness of 1st serve (E)	Probability 2nd serve in (A')	Effectiveness of 2nd serve (E')	Probability point won
.45	.85	.7	.4	.5365
.5	.8	.7	.4	.54
.3	.95	.7	.5	.53
.35	.9	.7	.5	.5425
.4	.8	.7	.5	.53
.35	.95	.8	.4	.54
.4	.85	.8	.4	.532
.45	.8	.8	.4	.536
.25	.95	.8	.5	.5375
.3	.85	.8	.5	.535
.4	.75	.8	.5	.54
.2	.85	.9	.5	.53
.4	.65	.9	.5	.53
.5	.95	.7	.4	.615
.6	.85	.7	.4	.622
.5	.85	.7	.5	.6
.6	.8	.7	.5	.62
.45	.95	.8	.4	.6035
.5	.9	.8	.4	.61
.4	.95	.8	.5	.62
.5	.85	.8	.5	.625
.35	.9	.9	.5	.6075
.5	.8	.9	.5	.625
.65	.95	.7	.4	.715
.6	.95	.7	.5	.71
.65	.9	.7	.5	.7075
.65	.95	.8	.4	.697
.7	.9	.8	.4	.726
.55	.95	.8	.5	.7025
.6	.9	.8	.5	.7
.5	.95	.9	.5	.7
.55	.9	.9	.5	.6975

her first serve stroke in more often than the second serve stroke, the player should use it all the time. Thus the conditions $A < A'$ and $E > E'$ can be added to the analysis.

To assess the impact of hitting the second serve stroke as a first serve, adjust the formula by using A' and E' instead of A and E. This gives a probability of winning a point of $(A'E') + ((1 - A')A'E')$. Mathematically, testing the pertinence of the commentator's observation reduces to answering the question, "Is $(A'E') + ((1 - A')A'E')$ bigger than $(AE) + ((1 - A)A'E')$?". From match data, values for A, E, A', and E' can be computed, and thus the efficacy of the strategy can be assessed. By taking values for A', E', and E, and solving $(A'E') + ((1 - A')A'E') = (AE) + ((1 - A)A'E')$ for A, we create the data for Table 20.6. Table 20.6 tabulates values for A' and E' and threshold values for A and E for which any larger efficiency or percentage give a higher expected return for hitting the first serve stroke, then the second serve stroke (if necessary) than for hitting the second serve stroke as a first serve.

The numbers in Table 20.6 contain information that runs counter to the traditional wisdom. If a player has a second serve that goes in 90% of the time and is 40% effective, the player only has to get a serve that wins three of four points in one time in ten to improve her results. What's more, the chart contains break-even values; increase either the efficiency or the percentage in of the first serve values and the likelihood of winning a point increases. Thus, even though hitting a 90% in, 60% effective stroke twice will give the same expectation for winning a point as a 15% in, 90% effective serve backed by a 90% in, 60% effective second serve, increasing to 30% in, 90% effective further increases the expectation of winning the point. If one considers the above scenario, that is, increasing the percentage in on a 90% effective first serve being backed by the 90-60 second serve hypothesized translates into winning 83% of the service games. And, if one could get half his first serves in and maintain the 90% effectiveness, the number soars to an astounding 98% of the games. (These numbers were computed using the numbers in Table 20.1, which relates point probabilities to game probabilities.) If further encouragement were needed, the "percentage in" factor can be accurately gauged on the practice court, although the effectiveness factor is likely dependent on the opponent.

The numbers speak quite clearly about service. If a player can get her second serve in, it makes little sense to use the second serve stroke as a first serve. If a player can't get a second serve in, the player shouldn't be worrying about tactics, he should be hitting buckets of serves.

Table 20.6.

2nd serve percentage (A')	2nd serve efficiency (E')	For 1st serve efficiency of	1st serve % must be at least
.7	.4	1	.12
		.84	.15
		.73	.18
		.67	.22
		.5	.38
.7	.5	1	.22
		.82	.32
		.75	.38
		.7	.45
.8	.4	1	.22
		.82	.13
		.72	.16
		.6	.23
		.5	.36
.8	.6	1	.18
		.92	.22
		.82	.28
		.75	.35
		.71	.41
.9	.4	1	.06
		.76	.09
		.59	.16
		.5	.25
.9	.6	1	.12
		.9	.15
		.83	.18
		.75	.25
		.7	.35

20.6 Afterthoughts

The folk wisdom associated with a subject quite often turns out to be based in fact. The experiment discussed in the previous pages of looking at tennis through its numbers gives a window through which to examine some of the perceived truisms of the game.

Why is there such a big emphasis on playing the "big points" well? Our analysis indicates that, for a match to be close, it is extremely likely that the number of points won by each player will be close. Perhaps the "big points" are magnified in the players' minds because they are actually the margin of victory. Put those points anywhere else in the match and the outcome is not likely to change. For example, consider a deuce game that Player 1 wins. Let P_i stand for Player 1 winning point number i and O_i stand for Player 1's opponent winning point number i. Schematically, a game might look like

$$P_1 \ P_2 \ P_3 \ O_4 \ O_5 \ O_6 \ P_7 \ P_8.$$

Now make your own game. You may fill in the first six spaces with three P's and three O's any way you like, but P_7 and P_8 must fill in the last two. Now exchange P_8 with any of the O's (that is, put the "big point" outcome in a non-"big point" place) and look at the resulting game. Player 1 won the game without going to deuce. Serve changes and the players go to a different set of performance parameters without Player 1 having to face a deuce. The argument might be made that, in the context of the match, this may have made it an even bigger point. To argue that "big point" performance is a factor, one would need to show that if the outcome on the perceived "big points" were turned around, then the opponent wouldn't have had the edge in points, and that if the "big points" were placed elsewhere in the match under similar parameters, then the outcome would have been different.

The "experts" often harp about the intangibles affecting outcomes. Things like "mental toughness" and "killer instinct" often come up in discussions of winning the "big points." Appropriately enough, the model that has been built in this paper can allow testing for intangibles. If actual set scores (or scores in "big" shot games, possibly games on serve if serve is a "big" shot) routinely outperform what the observed point probabilities predict, then perhaps we could explain the phenomenon through intangibles. In the matches I have charted in which players have outperformed their numbers (and the great majority of matches go exactly as the numbers say they should), a breakdown into service games has removed the anomaly. I also suspect that a "big" shot analysis would be telling in practically any scenario in which a player outperforms her numbers. It would also be interesting to see if a player's reputed strengths were those that were statistically indicated.

Bill Tilden is supposed to have said "Never change a winning game. Always change a losing game." He apparently sensed what the analysis in this paper indicates. The analyses from Table 20.1 and Table 20.2 show that winning points wins matches and the "big" shot analysis shows that the addition of a weapon can change an otherwise losing game into a winning game.

This discussion is not intended to be exhaustive, but rather to indicate that tools to decide what parts of traditional wisdom are actually wisdom and what parts are simply wishful thinking are available. It appears to be reasonable to understand the tangibles of a player's game before putting its fate in the hands of intangibles.

Reference

[1] Vic Braden and Bill Bruns, *Vic Braden's Tennis for the Future*, Little, Brown (Boston), 1977.

About the Author

G. Edgar Parker received his BS degree in Mathematics with a minor in Philosophy and Religion from Guilford College in 1969 and his PhD in mathematics from Emory University in 1977. In a teaching career that now spans 40 years, Ed has taught in high school, junior college, an open-admissions university, and selective universities, and remains committed to Inquiry-Based Learning. Although the bulk of his research has been done on differential equations, sports (in particular baseball) are his great recreational passion, and he is often led to think about sports due to listening to or reading remarks from the media or coaches as they analyze games he has seen. Ed began playing tennis in 1977 and it continues to be his main competitive outlet.

Part IX

Track and Field

CHAPTER 21

The Effects of Wind and Altitude in the 400m Sprint with Various IAAF Track Geometries

Vanessa Alday and Michael Frantz

Abstract

We investigate the effects that wind and altitude have on the 400m sprint when run on various IAAF track geometries, with the work based on the senior project written by Vanessa and supervised by Michael. We validate Quinn's ordinary differential equations model using data from the 1999 World Athletics Championships. The model is based on Newton's Law for the energy balance of a runner, and Maple is used to solve the model's equations numerically. We confirm some non-intuitive results about the effect of a constant wind blowing from a fixed direction, and we modify the model to predict wind-assisted performances on both an equal quadrant track and a track from the ancient Greek games. Comparing the tracks provides information about the effects on performances on different standard tracks. We find performance differences between running lanes, indicating possible disadvantages of running in certain lanes. We find that the effect of altitude is significant but of little consequence with respect to differences in track geometry.

21.1   Introduction and an Early Model

Track and field meets include many events, among them the 400m sprint. In a standard International Association of Athletics Federations (IAAF) track, there are eight lanes, and a maximum of eight runners in a race. Although each IAAF track has the same dimensions, questions have arisen as to the effect that wind and altitude have on the runners' performances, regardless of the event. Several models have been created to describe their effects on the 100m sprint, the 200m sprint, and the 4 × 100m relay. Modeling these performances proved to be relatively simple, but until 2004 no one had ever tried modeling a 400m sprint because of the difficulty of accounting for two straights and two bends. In 2004 Mike Quinn published a model for the 400m sprint, using data from the 1999 World Athletics Championships. It (Quinn, 2004) is the basis for our model. Physical intuition suggests that a constant wind blowing across a closed loop track would have a detrimental effect on a runner's time regardless of the wind direction.

The basis of most mathematical models dealing with the effects of wind on sprinting performances traces back to the work of Archibald V. Hill, a British physiologist and biophysicist, and Joseph B. Keller, Professor of Mathematics and Mechanical Engineering at Stanford University. Hill was the first to provide a model for the energy balance of a runner by using Newton's Law (Alvarez-Ramirez, 2002). His work went unpublished and forty-six years later Keller would use the ideas that Hill introduced in 1927. Keller's equation of motion of a runner was:

$$\frac{dv(t)}{dt} = f(t) - \frac{1}{t}v,$$

where $v(t)$ is the runner's velocity at time t in the direction of motion, and $f(t)$ is the runner's total propulsive force per unit mass, which drives the runner forward and overcomes both the internal and external resistive force $\frac{v}{\tau}$ per unit mass (Keller, 1973). Keller assumed that the resistance is a linear function of v and that the damping coefficient τ is a constant. The works by Hill and Keller became known as the Hill-Keller model. One consequence of the Hill-Keller model is the prediction of the existence of a lower velocity limit that can be maintained indefinitely, that is, a running pace which could be maintained over an infinite time interval, which is not possible from a physiological viewpoint.

21.2 Quinn's Model

In 2003 Quinn extended Keller's model, as others had before him, to determine the effects of wind and altitude in the 200m sprint. Quinn's model included the reaction time of the sprinter, as well as air resistance. The reaction time for a world-class sprinter is rarely below 0.13s and averages about 0.15s. Reaction times also vary by gender, with men having a slightly faster reaction time than women (Quinn, 2003). Quinn's extended model equation is:

$$\frac{dv}{dt} = Fe^{-\beta t} - \frac{v}{\tau} - \alpha(v - v_w)^2,$$

where $v = \frac{ds}{dt}$, v_w is the velocity of the wind relative to the ground and tangent to the path, and $\alpha = \frac{\rho C_d A}{2M}$ where ρ is the air density, taken to be 1.184 kg/m at 25°C. C_d is the coefficient of drag, taken to be 0.715 (Walpert and Kyle, 1989), A is the frontal area of the athlete, estimated to be 0.51m^2 for men (Quinn, 2004), and M is the mass of the athlete, taken to be 76 kg, a typical mass for a world class 400m sprinter. Quinn replaced $f(t)$ by $Fe^{-\beta t}$, a propulsive force per unit mass that diminishes during the race as the athlete's muscles tire.

21.3 The Effects of Track Geometry on Running Performance

The International Association of Athletics Federations track, otherwise known as the standard running track, is a 400m track measured along lane 1, with two straights each 84.4m long and two bends each 115.6m long. Thus 57.8% of the 400m event is run around bends. Although each runner runs around both bends, they experience different conditions in different lanes. Lanes have different radii around bends, though the length of the straight is the same for all. Thus runners will experience different wind conditions and the maximum velocity that runners can obtain around bends varies, depending on the lane.

The maximum velocity in running around a bend is less than that obtained while running in a straight line. Greene (1985) described the effects that runners experience around bends. He pointed out that lanes are unequal because of the effect of their radii on the runners' speed, since in order to balance centrifugal acceleration, a runner must heel over into the turn, with the approximate centerline of his body making an angle θ with respect to the vertical.

The final result of Greene's analysis is that everywhere the relation of the runner's peak velocity v_0 to his velocity v in a bend of radius r is $v = v_0 \sqrt{\omega}$, where ω satisfies the cubic equation $\omega^3 + r^2\omega - r^2 = 0$, $r = \frac{Rg}{v_0^2}$, R is the bend radius, g is the gravitational force, and v_0 is the runner's peak velocity. The cubic has a real root,

$$\omega = \left(\frac{\lambda^2}{2} + \sqrt{\frac{\lambda^4}{4} + \frac{\lambda^6}{27}} \right)^{1/3} - \left(-\frac{\lambda^2}{2} + \sqrt{\frac{\lambda^4}{4} + \frac{\lambda^6}{27}} \right)^{1/3}$$

Table 21.1. Lane radii for the IAAF standard track in meters

Lane	Radius (in meters)
1	36.80
2	37.92
3	39.14
4	40.36
5	41.58
6	42.80
7	44.02
8	45.24

where $\lambda = rg/v_o^2$. This may be one of the more unexpected and interesting appearances of Cardano's formula for obtaining the exact solution of a cubic polynomial equation! From Ward-Smith and Radford (2002) we obtain the following radii for the eight lanes, shown in Table 21.1.

21.4   Computation of the Effect of Winds

Another parameter in Quinn's model is v_w, the wind velocity relative to the ground and tangent to the path of the runner. It is different in each of the four segments of the track, and depends on the wind velocity u_w and the relative wind direction, which will vary continuously as the runners progress around the track. For each runner to run exactly 400m, they are staggered at the starting line. The runner in lane 1 starts at the starting line, and will run 115.6m around the first bend before entering the back straight. The runner in lane 8, however, is the farthest from the starting line, and will run 89.1m around the first bend before entering the back straight. All runners from lanes 1 through 8 run the back straight for the entire 84.4m before entering the second bend. Once again, each runner runs a different distance around this second bend. The runner in lane 1 will once again run 115.6m around the second bend before entering the finishing straight. The runner in lane 8, however, will run 142.1m around the second bend before entering the finishing straight. Finally, all runners from lanes 1 through 8 will again run the finishing straight for the remaining 84.4m before crossing the finishing line. We can calculate the wind facing the runner in each of the eight lanes.

We determine the wind velocity v_w as a function of $s(t)$, the distance traveled around the first bend. The component of the wind blowing in the direction of the runner for a runner traveling in a straight line with a wind velocity of u_w blowing at an angle θ, is $v_w = u_w \cos(\theta)$, where θ is the angle measured counterclockwise from the finishing straight on the lower edge of the track, as illustrated in Figure 21.1.

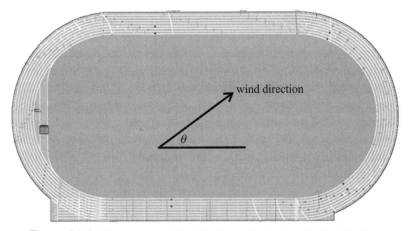

Figure 21.1. The geometry of the IAAF standard track and wind direction

Table 21.2. Relative wind velocity around the IAAF standard track

Relative Wind Velocity v_w	Region of Application
$v_w = u_w \cos\left(\theta + \dfrac{231.1 - s}{r_i}\right)$	First bend: $0 \leq s \leq 231.2 - \pi r_i$
$v_w = -u_w \cos\theta$	Back straight: $231.2 - \pi r_i \leq s \leq 315.6 - \pi r_i$
$v_w = u_w \cos\left(\theta + \dfrac{315.6 - s}{r_i}\right)$	Second bend: $315.6 - \pi r_i \leq s \leq 315.6$
$v_w = u_w \cos\theta$	Finishing straight: $315.6 \leq s \leq 400$

Trigonometry leads to expressions for v_w in each of the four regions of the track, for lane i with corresponding radius r_i as a function of distance traveled by the runner. Table 21.2 lists the four formulas and the domains where they apply.

21.5   Altitude and the Propulsive Force

We now consider the propulsive force, $Fe^{-\beta t}$. Although the drag coefficient is not affected by an increase in altitude, and variations in the gravitational acceleration due to altitude are negligible (Behncke, 1994), McArdle (2007) observed that increasing altitude forces a drop in the partial pressure of oxygen, reducing the percentage of oxygen saturation in the blood. This in turn decreases the oxygen supplied to the runner's muscles by the aerobic energy system. The oxygen saturation reduction is relatively small for altitudes less than 3000 meters, but still has a detrimental effect on athletic performances (Quinn, 2004). The conclusion is that the aerobic energy system contributes significantly to the energy supply during long sprints and middle distance running. The total contribution of the aerobic energy system for the 400m event was found to be 43% (Spencer and Gastin, 2001). Since the aerobic energy system is affected by altitude, the propulsive force $Fe^{-\beta t}$ declines at a faster rate in higher altitudes than at sea level. We assume that the decay rate β depends on both the altitude H and the contribution of the aerobic energy system γ and assume that $\beta = \beta_0(1 + \gamma\sigma H)$, where β_0 is the parameter value at sea level and σ is the oxygen saturation in the blood. Using estimates from McArdle (2007), we set $\beta_0 = 6.00 \times 10^{-3}s^{-1}$, $\gamma = 0.35$, and $\sigma = 0.000023\text{m}^{-1}$.

21.6 Data Collected and Results from Quinn

In Quinn's 2004 model, he chose the maximum force F, the decay rate β, and the resistive force τ to fit the available 400m data from Ferro's 2001 paper. The data provided in Ferro's analysis of the 400m event were produced using twelve analog and digital video cameras. Quinn was able to fit his model to the data by setting $F = 7.91$ m/s^2, $\beta = 6 \times 10^{-3}$ 1/s, and $\tau = 1.45s$ (for men).

Using the same ordinary differential equations and the same parameters as Quinn, a program written in *Maple* to solve the system of differential equations numerically replicated Quinn's model results. Table 21.3 shows the men's 400m simulation in lane 4 from Quinn's paper and compares his model time with the data for each of the 50m intervals. We also include results from the *Maple* program. A reaction time of 0.15 seconds was added to determine the final finishing time, and we assumed windless conditions.

Our model times closely approximate Quinn's results, the major difference being 0.04 seconds at 150m. The finishing time is the same for all three models.

21.7 Effects of Wind Direction on Overall Performance

We next consider the effects that wind has on the overall performance of a runner for different wind direction angles θ. Table 21.4 illustrates the effects for runners in lanes 1, 4, and 8 of a 2 m/s wind blowing from all points of the compass in 30° increments. (We use 2 m/s because it is the maximum legal speed for a world record to be recorded without the notation "wind assisted.") In addition to the predicted model times, we have included time

Table 21.3. 400m simulation in lane 4 for world-class runners

Distance (m)	Actual Time (s)	Quinn's Model Time (s)	Alday-Frantz Time (s)
50	5.99	6.07	6.10
100	10.95	10.99	10.99
150	15.95	15.93	15.89
200	21.07	20.98	20.95
250	26.27	26.29	26.27
300	31.51	31.75	31.76
350	37.03	37.33	37.32
400	43.03	43.03	43.03
Finish time (s)	43.18	43.18	43.18

Table 21.4. Effect of a 2 m/s wind on lanes 1, 4, and 8, with time corrections

Wind Dir. ($\theta°$)	Lane 1 (43.15)		Lane 4 (43.03)		Lane 8 (42.89)	
	Time (s)	Correction	Time	Correction	Time	Correction
0	43.23	+0.08	43.13	+0.10	43.03	+0.14
30	43.22	+0.07	43.13	+0.10	43.04	+0.15
60	43.19	+0.04	43.09	+0.06	43.01	+0.12
90	43.18	+0.03	43.06	+0.03	42.97	+0.08
120	43.17	+0.02	43.04	+0.01	42.92	+0.03
150	43.18	+0.03	43.03	0.00	42.88	−0.01
180	43.19	+0.04	43.02	−0.01	42.85	−0.04
210	43.18	+0.03	43.01	−0.02	42.82	−0.07
240	43.17	+0.02	43.02	−0.01	42.82	−0.07
270	43.17	+0.02	43.03	0.00	42.86	−0.03
300	43.19	+0.04	43.07	+0.04	42.92	+0.03
330	43.21	+0.06	43.11	+0.08	42.99	+0.10

corrections from the data obtained under windless conditions. For example, a runner in lane 4 would have an overall increase in time of 0.06 seconds corresponding to a steady 2 m/s wind blowing at an angle of 60° from the main straight.

According to these data, for a runner running in lane 4 the most favorable direction is $\theta = 210°$, that is, a wind blowing toward the southwest, where there is a slight advantage over running in windless conditions. A wind angle range anywhere between 180° and 240° also provides a slight advantage. For lane 8, the advantage is obtained when the wind angle lies between 150° and 270° degrees, with an optimal reduction of 0.07 seconds in the 210°–240° range. Lane 1 realizes the least detrimental effect when the wind blows at an angle from either 120° or 240°–270°, although the overall time is slower for all wind angles. In windless conditions, the time differential between lane 1 and lane 8 is 0.26 seconds, with the advantage going to the outer lane. This is a consequence of the slower velocities due to the smaller bend radius for the inner lanes, as modeled by Greene. We consider the radius effect with the interaction in the table.

The same data is displayed in Figure 21.2, with smoothed curves joining the data points. It is clear that for each lane there are two significant directions, with roughly opposite compass points, where there is a clear advantage or disadvantage if the wind blows in one of the directions.

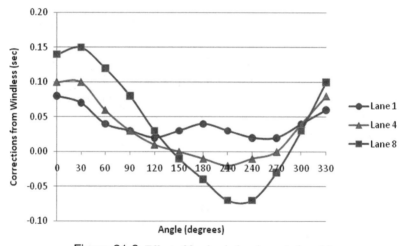

Figure 21.2. Effect of 2 m/s wind on lanes 1, 4, and 8

Although lane 8 seems to be the most favorable lane in windless conditions as well as in 2 m/s winds, most athletes prefer lanes 3, 4, 5, and 6. In championships these are allocated to the top-speed runners. The reasons why most athletes prefer these lanes are mostly strategic. Most runners prefer to see the progress of the other runners so they can pace their own effort. They are most visible from lanes 1 and 2, but the tighter bends incur a greater disadvantage than the strategic advantage. A runner's goal is to beat competitors rather than to achieve the shortest time, and that influences strategy (for example, recall Usain Bolt's 100m Olympic performance in 2008, when he slowed down significantly just before the finish line but still won the race).

21.8 Effects of Altitude and Air Density

Altitude has an effect on the performance of an athlete because of its effect on the aerobic energy system. There is a second more important factor, the air density. Races run at higher altitudes favor runners because air density is less at higher altitudes. We replace air density ρ in the expression for α, $\alpha = \rho C_d A / 2M$ with the air density ρ_H, at an altitude of H meters above sea level, which is related to the air density at sea level ρ_0 by $\rho_H = \rho_0 e^{gH/R(T+273)}$, where T is the air temperature in degrees Celsius, and the gas constant is $R = 287 J \cdot kg^{-1} \cdot K^{-1}$ (in joules per kilogram per degree Kelvin). In our model, $\rho_0 = 1.184$ kg/m^3 and $T = 25°C$, and the altitude, H, changes according to the model. Table 21.5 contains the time corrections for lanes 1, 4, and 8 on the standard track at various altitudes and under windless conditions.

Table 21.5. Time corrections for lanes 1, 4, and 8 for different altitudes in windless conditions

Altitude	Lane 1 (43.15)		Lane 4 (43.03)		Lane 8 (42.89)	
(m)	Time (s)	Correction	Time	Correction	Time	Correction
0	43.15	0	43.03	0	42.89	0
500	43.08	−0.07	42.96	−0.07	42.82	−0.07
1000	43.01	−0.14	42.88	−0.15	42.75	−0.14
1500	42.95	−0.20	42.83	−0.20	42.69	−0.20
2000	42.89	−0.26	42.77	−0.26	42.63	−0.26
2500	42.84	−0.31	42.71	−0.32	42.58	−0.31

The time corrections are almost identical across the lanes at each altitude. This to be expected, as the only effect on race times due to lane geometry comes from the physical mechanics of negotiation of tighter bends, which should not be affected by altitude. The largest time correction was a significant 0.32 seconds, which can easily make a difference in whether a world record is broken or not, and is why it has been accepted that altitude-assisted performances should be noted for races at altitudes higher than 1000 meters, and, according to the IAAF, these performances are marked by including an "A" notation.

21.9 The Equal Quadrant Track

Although the IAAF standard track is the norm for track dimensions, the IAAF accepts other tracks. One type of track is the equal quadrant track, which is a 400m track with 100m bends and 100m along each straightaway, measured along lane 1. The other type of track is the non-equal quadrant track, which is a 400m track, measured along lane 1, with two curved ends of equal radius and two straights equal in length but longer or shorter than the bends. Finally, there is a double-bend track, which is also a 400m track measured along lane 1, with two straights of equal length and two curves that are formed with different radii for each bend. Our model has only been applied to the IAAF standard track, but we will now examine how the wind affects a runner's performance on an equal quadrant track, illustrated in Figure 21.3.

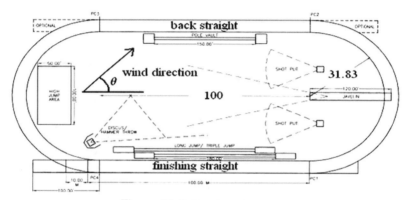

Figure 21.3. Equal quadrant track

We choose the equal quadrant track because before the IAAF standard track became so popular, the equal quadrant track was preferred. However, today the standard track is by far the most widely used design for a number of reasons, including a wider turning radius that favors runners and enhances performances, lessens injury, and allows greater flexibility in placing fields, especially soccer pitches, within the track oval.

Since the equal quadrant track differs from the IAAF standard track only in its dimensions, all the equations will remain the same, except for the wind velocity. Changes to the wind velocity functions can be easily determined, and the simulation runs with virtually the same code. Using the standard track data from Ward-Smith and Radford (2002), but now for an equal quadrant track, we have the lane radii in Table 21.6.

Analyzing as before, we obtain in Table 21.7 new expressions on different portions of the track for the relative wind velocity v_w.

Table 21.6. Turning radii for the lanes on an equal quadrant track

Lane	Radius (in meters)
1	31.83
2	32.95
3	34.17
4	35.39
5	36.61
6	37.83
7	39.05
8	40.27

Table 21.7. Relative wind velocity around the IAAF equal quadrant track

Relative Wind Velocity v_w	Region of Application
$v_w = u_w \cos\left(\theta + \dfrac{200 - s}{r_i}\right)$	First bend: $0 \le s \le 200 - \pi r_i$
$v_w = -u_w \cos\theta$	Back straight: $200 - \pi r_i \le s \le 300 - \pi r_i$
$v_w = u_w \cos\left(\theta + \dfrac{300 - s}{r_i}\right)$	Second bend: $300 - \pi r_i \le s \le 300$
$v_w = u_w \cos\theta$	Main straight: $300 \le s \le 400$

21.10 Wind Effects on the Equal Quadrant Track

Using the same equations, except with a modified definition and domain for v_w, we are able to produce model results for an equal quadrant track. Once again the finish time is increased by the 0.15 second reaction time. Table 21.8 shows the results of a 400m race in lane 4 for an equal quadrant track with no wind.

Table 21.8. Comparison of times for equal quadrant and standard tracks (windless)

Distance (m)	Equal Quadrant Model Time (s)	IAAF Standard Model Time (s)
50	6.13	6.10
100	11.02	10.99
150	15.91	15.89
200	20.98	20.95
250	26.34	26.70
300	31.86	31.76
350	37.39	37.32
400	43.09	43.03
Finish Time	43.24	43.18

Table 21.9. Effect of a 2 m/s wind on lanes 1, 4, and 8, with time corrections from windless

Wind Dir.	Lane 1 (43.25)		Lane 4 (43.09)		Lane 8 (42.93)	
($\theta°$)	Time (s)	Correction	Time	Correction	Time	Correction
0	43.33	+0.08	43.20	+0.11	43.07	+0.14
30	43.31	+0.06	43.19	+0.1	43.07	+0.14
60	43.29	+0.04	43.16	+0.07	43.05	+0.12
90	43.27	+0.02	43.12	+0.03	43.00	+0.07
120	43.27	+0.02	43.10	+0.01	42.96	+0.03
150	43.28	+0.03	43.09	0	42.93	0
180	43.28	+0.03	43.09	0	42.89	−0.04
210	43.28	+0.03	43.08	−0.01	42.87	−0.06
240	43.26	+0.01	43.08	−0.01	42.86	−0.07
270	43.27	+0.02	43.10	+0.01	42.90	−0.03
300	43.28	+0.03	43.14	+0.05	42.96	+0.03
330	43.31	+0.06	43.18	+0.09	43.03	+0.10

The final time in the equal quadrant track is 0.06 seconds slower than the IAAF standard track simulation, due to the smaller radii on the bends. This causes the runners to spend more time on the ground than in the air, which increases their time. An equal quadrant track has longer straights than the IAAF standard track, which gives the runners on an equal quadrant track an advantage to be able to run faster on the straights. However, this does not outweigh the disadvantage that they have on the bends.

Now that the model has been modified for an equal quadrant track, the effects of the direction of a 2 m/s wind can be computed, and are displayed in Table 21.9 for lanes 1, 4, and 8, in terms of race times and corrections relative to the windless time of 43.24 seconds.

The same data are displayed in Figure 21.4, smoothed.

The results are similar to those for the standard track. The optimal wind direction for lane 4 is still in the 210°–240° range, for lane 8 at 240°, and lane 1 remains disadvantaged at all wind angles, although the best race times are achieved at angles of either around 120° or 240°.

21.11 The Ancient Greek Olympiad Track

One of the purposes of mathematical modeling is to predict results that are either too expensive, too dangerous, or simply impossible to achieve by testing. We decided to apply our working model to another track configuration,

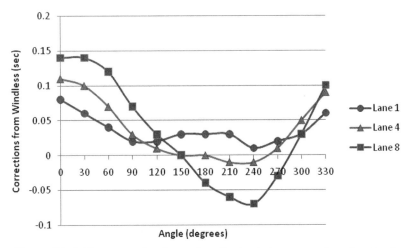

Figure 21.4. Time correction factors for a 2 m/s wind on an equal quadrant track

the track geometry used by the ancient Greeks in the early Olympic Games, and revived in the restoration of the Panathenaic Stadium in Athens in 1895 for the modern Olympics (see Figure 21.5). Our model can answer questions about how much time differential between performances on an ancient track and a modern track can be attributed to the track geometry alone, dis-

Figure 21.5. The Panathenaic Stadium in Athens

regarding the many other variables such as fitness, training, equipment, track surface, etc.

A track with this geometry has two straights of 180 meters each and two bends of radius 6.37 meters, each contributing 20 meters. All that is required is a change of the parameters governing the strength of the relative wind speed as a function of the distance around the track, analogous to the adjustments made in moving from the standard track to the equal quadrant track. Table 21.10 provides a comparison of the times in windless conditions for lane 4 in the Panathenian track with the IAAF standard track.

Table 21.10. Comparison of lane 4 (windless) times; IAAF standard vs. Panathenian track

Distance in meters	Std. track time in sec	Pan. track time in sec	Std. time per 50 m	Pan. time per 50 m
50	6.10	6.15	6.10	6.15
100	10.99	10.91	4.89	4.76
150	15.89	15.79	4.90	4.88
200	20.95	21.06	5.06	5.27
250	26.70	26.69	5.75	5.63
300	31.76	32.04	5.06	5.35
350	37.32	37.57	5.56	5.53
400	43.03	43.28	5.71	5.71
Finish time (s)	43.18	43.43	43.18	43.43

As might be expected, although the longer straights on the Panathenian track permit more of the race to be run at peak speed, the extremely sharp bends cause slowdowns that more than compensate, leading to an increase in race time of 0.25 seconds. Table 21.11 shows the effect of a 2 m/s wind on lanes 1, 4, and 8 on the Panathenian track.

We plot the correction factors induced by the wind, smoothed, in Figure 21.6.

21.12 Summary of Results

A comparison of results from the IAAF standard track and the equal quadrant track shows similarities as well as differences. The biggest disadvantages to the runner on both tracks in windless conditions are in lane 1, since the tighter bends force the runner to spend more time on the ground and less time in the air. The effect is greater for the equal quadrant track, since the radius of lane 1 is 31.83m, instead of the 36.80m radius of lane 1 for the IAAF

Table 21.11. Effect of a 2 m/s wind on lanes 1, 4, and 8, with time corrections from windless (Panathenian track)

Wind Dir. ($\theta°$)	Lane 1 (43.84)		Lane 4 (43.28)		Lane 8 (43.00)	
	Time (s)	Correction	Time	Correction	Time	Correction
0	43.93	0.09	43.38	0.10	43.08	0.08
30	43.91	0.07	43.38	0.10	43.09	0.09
60	43.87	0.03	43.35	0.07	43.08	0.08
90	43.85	0.01	43.32	0.04	43.06	0.06
120	43.86	0.02	43.31	0.03	43.07	0.07
150	43.88	0.04	43.32	0.04	43.08	0.08
180	43.90	0.06	43.31	0.03	43.06	0.06
210	43.80	0.04	43.28	0.00	43.02	0.02
240	43.86	0.02	43.25	−0.03	42.97	−0.03
270	43.85	0.01	43.25	−0.03	42.95	−0.05
300	43.87	0.03	43.29	0.01	42.97	−0.03
330	43.91	0.07	43.35	0.07	43.03	0.03

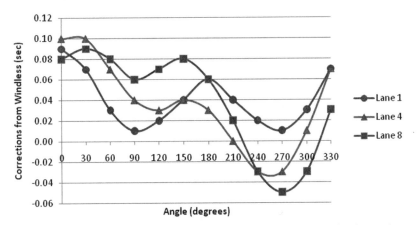

Figure 21.6. Time correction factors for a 2 m/s wind on the Panathenian track

standard track. Lane 8 appears to be the fastest lane on both tracks, having a shorter distance around the first bend than any other lane, and a wider radius around the second bend, permitting the runner to stay in the air longer than the other runners and thus producing a faster time than in any other lane. Although the radius of lane 8 on an equal quadrant track at 40.27m is less than that of the 45.25m radius of lane 8 on the standard track, which would seem to give an advantage to the standard track, the disadvantage of the standard track is that it has much shorter straights. On the standard track each straight is run for 84.4 meters, whereas on the equal quadrant track each straight is run for 100 meters. The more gradual (faster) bends of the standard track overcome the disadvantage of its shorter straights, providing a time that is 0.06 seconds faster in lane 4 under windless conditions. Although 0.06 seconds seems like a rather brief interval in many applications, it can feel like an eternity in a sprint race.

With a constant 2 m/s wind, the results indicate advantages and disadvantages. Lane 1, for example, offered no advantages from any wind direction, and a loss of 0.08 seconds on either track with a 0° headwind on the main straight. There is a slight advantage in lane 4 with a wind direction varying from 210° to 240°, providing a time correction of −0.02 and −0.01 seconds on the standard and equal quadrant tracks, respectively. Lesser advantages occur on both tracks with wind directions approximately between 180° and 250°. The largest wind advantage is found in lane 8 on each track, with a wind direction from 210° to 240° providing a time correction of −0.07 seconds. Lesser time advantages occur in lane 8 with a wind direction approximately between 150° and 280°.

Since the Panathenian track is no longer in competitive use, our interest in examining it was simply to observe what kind of advantages or disadvantages might be predictable from our model, and whether or not they have any relationship to what one might predict from the physics and geometry of the situation. The amount of time spent on the bends is minimal in comparison with the other two tracks, supporting an argument that reducing the headwind on the straights would serve the runner best. The model indicates an optimal time in all three lanes (1, 4, and 8) for a southerly wind of 270°, with another lesser local minimum time for a northerly, or 90° wind, both blowing perpendicular to the runner for most of the race. The corollary to this is that wind directions parallel to the straights, at about 0° or 180°, provide the worst conditions, as might be expected. The data illustrate a principle that holds for all three tracks, namely that the most benefit occurs for the runner facing a wind direction of 240°, while the worst outcome is associated with a wind direction of about 30°. One interpretation of this is that if a wind must be encountered, and particularly on a portion of the track which requires more energy (like the bends), it is advantageous to have the wind at the runner's back earlier in the race, and then to have the headwind later. If one considers two races that are run parallel to the wind, race A consisting of a mile with a tailwind, then a mile with a headwind, and race B with the headwind first and then the tailwind, runners in race A have an advantage over those in race B because the energy gain from a tailwind is known to be about half the size of the energy loss due to a headwind. This means that a headwind early in the race has a greater effect in decreasing a runner's energy for the remainder of the race, as opposed to the benefits reaped from an early tailwind.

As with the IAAF standard track, we tested the effects of altitude on the equal quadrant track and Panathenian track. The results were entirely consistent with the thesis that the higher the altitude, the faster the times. The average altitude time corrections were -0.09, -0.18, -0.27, -0.35, and -0.42 seconds for altitudes of 500 meters, 1000 meters, 1500 meters, 2000 meters, and 2500 meters.

We have shown that wind direction can significantly affect the performance of a runner in a 400m race. When sprinting, the disadvantage of a head wind is greater than the benefit of a tail wind of the same magnitude, so wind direction plays a role in the performance of a runner to the extent that a record may be set without wind assisted conditions on one day, only to be broken on the next day under a wind of the same velocity but from a different direction.

Finally, the effect of altitude on race times is also significant and predictable. Altitude has a major effect on the overall time for both tracks, the biggest time corrections being for the equal quadrant track. Our model, inspired by Quinn, shows that other important factors are at stake, including lane selection, wind speed and direction, location and altitude, and even track geometry. Running in track and field events and winning is not as simple as being in shape and generating a strategy.

21.13 Directions for Possible Future Work

Quinn's 2004 model considered winds of constant speed and direction throughout the stadium. The architecture of the stadium also affects how wind changes, depending on the placement of the bleachers, their height and width, and other factors that might produce erratic winds. Creating a model for these circumstances could result in more accurate outcomes, but there are many other variables not accounted for. In addition, improvements on the performance of a 400m run are not attributable solely to wind and altitude.

For example, lane 8 is the theoretical fastest lane, which indicates that it should be the favorite choice of top sprinters. Anyone who has watched a major track event knows that in fact the goal for most runners is to have a lane assignment in one of lanes 3 through 6. There are other strategic factors at work here, involving seeing more easily where the competition is, avoiding the curb adjacent to the inner track, being unable to see the competition at all, and being of a disposition which prefers leading to catching up, or vice versa.

Other factors exist that affect a runner's performance, including physical condition, nervous tension, health, clothing, personal issues, etc. Measuring and modeling them range from difficult to impossible. Another factor that has been overlooked, as Keller (1973) states, is that the goal of runners is to beat competitors rather than to achieve the shortest time, which influences their strategy.

Although many of these factors appear to be hopelessly non-quantifiable, it is the job of the applied mathematician to ferret out what the most significant factors are, and to discover a method to account for them in the model. Perhaps this work will inspire others to look for ways to improve it, or generate a completely new model, or turn mathematics to a description of other sports which have thus far been ignored, in the hope of obtaining a better understanding of all the factors that go into a world record performance.

References

[1] J. Alvarez-Ramirez, An improved Peronnet-Thibault mathematical model of human running performance, *European Journal of Applied Physiology* 86 (2002) 617–525.

[2] H. Behncke, Small effects in running, *Journal of Applied Biomechanics* 10 (1994) 270–290.

[3] A. Ferro, A. Rivera, and I. Pagola et al., Biomechanical research project at the 77th World Championships in Athletics Seville 1999, *New Studies in Athletics* 16 (2001) 25–60.

[4] P. R. Greene, Running on flat turns: experiments, theory, and applications, *Journal of Biomechanical Engineering: Transactions of the American Society of Mechanical Engineers* 107 (1985) 96–103.

[5] J. B. Keller, A theory of competitive running, *Physics Today* 26 (1973) 42–47.

[6] W. D. McArdle, F. I. Katch, and V. L. Katch, V.L., *Exercise Physiology: Energy, Nutrition and Human Performance,* 6th ed., Lippincott, Williams, and Wilkins, Philadelphia, PA, 2007.

[7] M. D. Quinn, The effects of wind and altitude in the 200m sprint, *Journal of Applied Biomechanics* 19 (2003) 49–59.

[8] M. D. Quinn, The effects of wind and altitude in the 400m sprint, *Journal of Sports Science* 22 (2004) 1073–1081.

[9] M. R. Spencer and P. B. Gastin, Energy system contribution during 200 to 1500m running in highly trained athletes, *Medicine and Science in Sports and Exercise* 33 (2001) 157–162.

[10] R. A. Walpert and C. R. Kyle, Aerodynamics of the human body in sports, *Journal of Biomechanics* 22 (1989) 1096.

[11] A. J. Ward-Smith and P.F. Radford, A mathematical analysis of the 4×100m relay, *Journal of Sports Sciences* 20 (2002) 369–381.

About the Authors

Vanessa Alday is a graduate student in the MBA program at California State University at Fullerton. She holds a bachelor's degree in mathematics from the University of La Verne.

Michael Frantz received his PhD in mathematics from the Claremont Graduate School. He taught at Wichita State University before moving to the University of La Verne, where he is now Professor of Mathematics and Chair of the Department of Mathematics, Physics, and Computer Science. He is currently involved in the mathematical modeling of problems concerned with the fluid dynamics in fuel cell technology, and also maintains a healthy interest in the interaction between mathematics, art, and music.

CHAPTER 22

Mathematical Ranking of the Division III Track and Field Conferences

Chris Fisette

Abstract

In this article we provide a mathematical model for ranking of the National Collegiate Athletic Association (NCAA) Division III track and field conferences. It uses four ranking systems involving vector lengths and z-scores to measure and rank the strength of each conference using the top eight marks of each conferences championship meet. We take an average of each conferences rank to produce the final ranking. We believe that this model is superior to the current ranking system at the national meet. We conclude by using the 2008 mens outdoor season data to rank the Division III conferences for the 2009 season.

In the National Collegiate Athletic Association (NCAA), there are no ranking systems for the track and field conferences in any division. The only rankings in track and field are the school rankings at the national championship meet. Because of the way they are made, the national champion school may not even be the strongest team in its conference.

At the national meet, it is possible to win the national title with only a few top athletes. In 2008, the outdoor national title was won with a total of only 35 points. These could have been earned with three first place finishes and one fourth place finish, so the national title can be won by a school that has a single exceptional athlete. In 2008 Fisk University was tied for 29th in Division III with 10 points. Fisk University's team consisted of a single athlete who won his event at the national meet. With the ranking method used at the national meet there are conferences that do not score any points, so

Fisk University's team of one athlete appears to be better than entire conferences. This is absurd. The school rankings at the national championship do not reflect the strength of a school or a conference. Rather, they show only how the top one or two athletes in each conference compare to each other. It is like ranking entire baseball teams on the basis of their best hitter and best pitcher.

People familiar with track and field form opinions about which conferences are the best based on rankings at the national meet. They assume that the conferences with the top ranking teams are the best conferences in the nation. However, if Conference A has a single All-American in every event while all the others in Conference A are mediocre while Conference B has many excellent athletes who are slightly below the level required to compete at the national championship meet, the ranking system at the national meet would make it appear that Conference A is stronger than Conference B even though were the two conferences to compete against each other in a meet, Conference A would win every event, but Conference B would easily win the meet.

This shows that the ranking method used at the national meet does not reflect the strength of the conferences. To rank conferences more accurately there needs to be a ranking system that takes into account the performances of the athletes at normal meets and not just at the national meet.

In what follows, I use vector lengths and z-scores to create a ranking system of the Division III track and field conferences that I believe is superior to the current system used at the national championship meet. Because athletes are usually at their best around the time of the conference meet I use their performances at the conference meet as data points.

For each event to have equal weight we need to normalize them by using ratios of a standard mark for an event and the average value for that event. In Division III there are two standard marks for each event, the national provisional mark and the national automatic mark. To qualify for the Division III National Championship Meet, athletes need to better the national provisional mark. The national championship takes a set number of athletes for the entire meet. They include all who have achieved the automatic mark and fill the remainder of the spots with athletes with the highest provisional marks. For our model we create a value without units that gives each event equal weight. Because the goal in track events is to achieve a time lower than the automatic mark, we compute the value by dividing the automatic mark by the average value for that event. For field events, the value is obtained by dividing the average value of each event by automatic mark because the goal is to exceed the automatic mark. Since these values are within a few hundredths of the ratio of automatic mark and the provisional mark they give each event equal weight.

In our model, two vectors, $\mathbf{Vr}$ and $\mathbf{Vz}$ each consisting of 20 values are employed. The vector $\mathbf{Vr}$ uses the ratios of the NCAA Division III automatic mark and the event averages. The vector $\mathbf{Vz}$ uses the event z-scores. The vector components in order are the ratios for the events: 100, 200, 400, 800, 1500, 5000, 10000, 110 hurdles, 400 hurdles, steeplechase, $4 * 100$, $4 * 400$, long jump, triple jump, shot put, discus, hammer, javelin, high jump, pole vault. In this model there are two ranking methods that use vector lengths. The length of the vector $(a_1, a_2, \ldots, a_{20})$ is defined as the square root of $a_1^2 + a_2^2 + \cdots + a_{20}^2$. A z-score is a measure of the distance in standard deviations of a sample mean to the population mean. That is, $z = (\bar{x} - \mu)/\sigma$, where $\bar{x}$ is the sample mean, μ is the population mean, and σ is the standard deviation. z-scores can be converted to probabilities using the standard normal curve. We are able to use z-scores since the results at championship meets are normally distributed.

For the second method of ranking conferences, we adapt the method that is used to score cross country meets. In cross country meets, the first place runner gets one point, the second place runner gets two points, and so on. To calculate the team score, points of the top five individuals of each team are added. The team with the lowest score wins the meet, the team with the second lowest score takes second, and so on. In cross country meets there are no ties. To break ties, the score of the teams' sixth runners are included. Whichever team's sixth runner finished higher takes the higher position. If a team has only four runners, it gets a point value for their fifth runner by using the score of the last place finisher in the meet plus one.

Our model assigns each event the same number of points and the points are totaled for each conference. As in cross country scoring, the conference with the lowest point value will be ranked number one, the conference with the second lowest point value will be ranked number two, and so on. If a conference has no entries in an event, then it gets the highest point value in the event. Since there are thirty-two conferences, every event will have the point values from one to thirty-two, so there is a total of 528 points for each event. If two conferences tie for nth place, we assign the average of n and $n + 1$ points to each. This keeps the number of performances throughout the conferences constant and the number of points in each event the same. If there is exactly one conference that does not have a certain event at their championship meet, that conference will be ranked last and assigned a point value of thirty two in that event. If there is more than one conference that does not have an event, we use the cross country method for breaking ties to assign point values. We call this method of counting the *modified cross country count*. We do not use the decathlon in our model because most conferences do not include it as a conference championship event.

We begin by averaging the top eight performances of all twenty events at each of the conferences' championship meet. Since twelve events are measured in units of time and eight are measured in units of distance, the averages need to be converted into a value that will give them a common unit and equal weight. It is important that the events have an equal weight to keep certain events from skewing the results. To do this, we use ratios involving the averages and the national automatic marks. Since lower times are better than the automatic mark in track events, we will divide the automatic marks by the averages for them. For field events, marks greater than the automatic mark are better so we will divide the averages by their respective automatic marks. This gives ratios with no units and equal weight that can be compared.

With these ratios, we make the vector $\mathbf{Vr}$ for each conference and calculate its length. We rank the conferences' vectors from longest to shortest with the conference having the longest vector length first. This gives us our first preliminary conference ranking.

We next use the z-scores and the modified cross country scoring method to assign points in each event and total them for each conference. Because faster times for running events will have a negative z-value and the slower times will have a positive z-value, we use the absolute value of the z-score in all our calculations. We use the modified z-scores to obtain the three further preliminary conference rankings. To get the overall ranking for each conference we total the points for each conference. The best conference is the one with the fewest points and the worst conference is the one with the most points. This gives us our second preliminary conference ranking.

The third ranking method ranks each conference by averaging the z-scores of each event within a conference. Ordered from largest to smallest, they give us a third preliminary conference ranking from best to worst.

For the fourth ranking method, we convert the z-scores into their corresponding probabilities from the normal curve. When a conference does not have an event, we assign that conference a probability of zero for that event. We then use these probabilities to create a twenty-component vector, $\mathbf{Vz}$, for each conference using the same event order as used for $\mathbf{Vr}$. Our fourth preliminary ranking of the conferences is by the lengths of the $\mathbf{Vz}$ vectors, from longest to shortest.

To produce our final conference ranking we average the four preliminary rankings. We conclude by applying our model to produce a ranking system of the NCAA Division III track and field conferences for the 2008 outdoor season. By employing four preliminary rankings, we were able to use a variety of methods to rank the conferences. Even though some of the conferences differed in their position from method to method, they were not far apart. In fact, there were two conferences whose position was exactly the same through every ranking. In the method that used the modified cross country count, nineteen of the thirty-two conferences had the same ranking as the final ranking, including the bottom eight.

Although some of the conferences near the top of our final ranking are also near the top of the national championship ranking there were some significant differences. This demonstrates that the scores at the national championship meet reflect only the strength of the top few athletes in a conference rather than the overall strength of the conference.

Our second preliminary ranking gives an added bonus. It ranks every conference by event which allows us to group certain events together to see where conferences rank based only on those events. This enables us to see which conference is the best sprinting conference, distance conference, field event conference, and so on. Also, it allows athletes who are curious as to where their conference ranks nationally based only on their events a way to see where their conference ranks by adding the points for each conference in just those events.

Our assumption in our model is that athletes have near peak performances in their conference championship. Weather conditions and other factors may affect an athlete's performance, as can things such as injuries, false starts, and disqualifications, especially for top athletes.

In the tables on the following pages conferences are identified by their usual abbreviations.

2008 Ranking of Division III Track and Field Conferences

z-Score Count			Ratio Vector Lengths			z-Score Averages		
1st	MIAC	1.00	1st	WIAC	1.00	1st	WIAC	1.00
2nd	WIAC	2.00	2nd	MIAC	2.00	2nd	MIAC	2.00
3rd	IIAC	3.00	3rd	IIAC	3.00	3rd	IIAC	3.00
4th	CCIW	4.00	4th	NESCAC	4.00	4th	CCIW	4.00
5th	NESCAC	5.00	5th	NWC	5.00	5th	NESCAC	5.00
6th	OAC	6.00	6th	OAC	6.00	6th	OAC	6.00
7th	UAA	7.00	7th	MWC	7.00	7th	SUNYAC	7.00
8th	SUNYAC	8.00	8th	SUNYAC	8.00	8th	NWC	8.00
9th	NWC	9.00	9th	UAA	9.00	9th	UAA	9.00
10th	MWC	10.00	10th	NEWMAC	10.00	10th	MWC	10.00
11th	MIAA	11.00	11th	CCIW	11.00	11th	SCIAC	11.00
12th	SCIAC	12.00	12th	CC	12.00	12th	MIAA	12.00
13th	CC	13.00	13th	MIAA	13.00	13th	CC	13.00
14th	NEWMAC	14.00	14th	NJAC	14.00	14th	NEWMAC	14.00
15th	NJAC	15.00	15th	MAC	15.00	15th	NJAC	15.00
16th	MAC	16.00	16th	SCIAC	16.00	16th	MAC	16.00
17th	ODAC	17.00	17th	HCAC	17.00	17th	ODAC	17.00
18th	PAC	18.00	18th	SCAC	18.00	18th	PAC	18.00
19th	HCAC	19.00	19th	LC	19.00	19th	HCAC	19.00
20th	LC	20.00	20th	PAC	20.00	20th	LC	20.00
21st	USAS	21.00	21st	USAS	21.00	21st	SCAC	21.00
22nd	NCAC	22.00	22nd	ODAC	22.00	22nd	USAS	22.00
23rd	ASC	23.00	23rd	CAC	23.00	23rd	Empire 8	23.00
24th	SCAC	24.00	24th	NAC	24.00	24th	ASC	24.00
25th	CAC	25.00	25th	LEC	25.00	25th	NCAC	25.00
26th	Empire 8	26.00	26th	MSCAC	26.00	26th	CAC	26.00
27th	LEC	27.00	27th	Empire 8	27.00	27th	LLC	27.00
28th	LLC	28.00	28th	LLC	28.00	28th	LEC	28.00
29th	NAC	29.00	29th	ASC	29.00	29th	NAC	29.00
30th	MSCAC	30.00	30th	UMAC	30.00	30th	MSCAC	30.00
31st	CUNYC	31.00	31st	NCAC	31.00	31st	UMAC	31.00
32nd	UMAC	32.00	32nd	CUNYC	32.00	32nd	CUNYC	32.00

z-Score Vector Lengths		
1st	WIAC	1.00
2nd	MIAC	2.00
3rd	IIAC	3.00
4th	CCIW	4.00
5th	NESCAC	5.00
6th	OAC	6.00
7th	SUNYAC	7.00
8th	NWC	8.00
9th	UAA	9.00
10th	MWC	10.00
11th	SCIAC	11.00
12th	MIAA	12.00
13th	NEWMAC	13.00
14th	NJAC	14.00
15th	CC	15.00
16th	MAC	16.00
17th	HCAC	17.00
18th	PAC	18.00
19th	ODAC	19.00
20th	NCAC	20.00
21st	LC	21.00
22nd	USAS	22.00
23rd	SCAC	23.00
24th	ASC	24.00
25th	Empire 8	25.00
26th	CAC	26.00
27th	LLC	27.00
28th	LEC	28.00
29th	NAC	29.00
30th	MSCAC	30.00
31st	CUNYC	31.00
32nd	UMAC	32.00

Final Ranking		
Rank	Conference	Average
1st	WIAC	1.25
2nd	MIAC	1.75
3rd	IIAC	3.00
4th	NESCAC	4.75
5th	CCIW	5.75
6th	OAC	6.00
7th	NWC	7.33
8th	SUNYAC	7.50
9th	UAA	8.50
10th	MWC	9.25
11th	MIAA	12.00
12th	SCIAC	12.50
13th	NEWMAC	12.75
14th	CC	13.25
15th	NJAC	14.50
16th	MAC	15.67
17th	HCAC	18.00
18th	PAC	18.50
19th	ODAC	18.75
20th	LC	20.00
T-21st	USAS	21.50
T-21st	SCAC	21.50
23rd	NCAC	24.50
24th	ASC	25.00
25th	CAC	25.00
26th	Empire 8	25.25
27th	LEC	27.00
28th	LLC	27.50
29th	NAC	27.75
30th	MSCAC	29.00
31st	CUNYC	31.25
32nd	UMAC	31.50

NCAA Championship Rank		
Rank	Conference	Points
1st	WIAC	97
2nd	IIAC	70
3rd	CCIW	54.5
4th	SUNYAC	54
5th	MIAC	52
6th	NESCAC	49
7th	NWC	42
8th	OAC	37
9th	ASC	35
10th	UAA	34
11th	USAS	27
12th	MAC	26
13th	MWC	25
14th	LC	24
T-15th	MIAA	23
T-15th	NJAC	23
17th	Empire 8	20
18th	PAC	19
19th	ODAC	18
20th	SCIAC	12
21st	SCIAC	11
22nd	HCAC	9.5
23rd	NCAC	9
24th	SCAC	8.5
25th	NAC	8
26th	NEWMAC	5.5
27th	CUNYC	4
28th	CC	3
29th	LEC	1
T-30th	CAC	0
T-30th	LLC	0
T-30th	UMAC	0

Acknowledgement   I wish to thank Joe Gallian for his assistance with the exposition in the article.

About the Author

Chris Fisette received a B.A. in mathematics in 2009 from Saint Mary's University of Minnesota in Winona, Minnesota. While attending college, he competed in four years of track and field. He is currently enrolled in the Graduate Teaching Licensure Program at the College of Saint Scholastica in Duluth, Minnesota, where he is also an assistant track and field coach.

CHAPTER 23

What is the Speed Limit for Men's 100 Meter Dash

Reza Noubary

Abstract

Sports provide an inexhaustible source of fascinating and challenging problems in many disciplines, including mathematics. In recent years, due to the emergence of some exceptional athletes, prediction of athletic records has received a great deal of attention. For example, mathematicians have tried to model the improvements of records over time in order to forecast future records, including ultimate records. Records set in different sports shed light on human strengths and limitations and provide data for scientific investigations, training, and treatment programs.

This article reviews some common methods used for modeling and analysis of athletic performances and the effect an exceptional individual like Usain Bolt could have on the results. Methods discussed include trend analysis, tail modeling, and methods based on certain results of the theory of records. Data from a few athletic events including the men's 100m dash are used for demonstration.

23.1 Introduction

At the August, 2009 world track and field competitions in Berlin, Usain Bolt, the Jamaican sprinting sensation put on some amazing performances, shattering his own records in both the 100 and 200 meter lowering both by 0.11 seconds to an amazing 9.58 seconds in the 100 meter and 19.19 in the 200 meter. The man is certainly on another level.

His time is the greatest improvement in the 100 meter record since electronic timing began in 1968. Bolt is not done yet and who knows how fast he can run. In fact, he thinks he can do better. He said this after winning the 100 meter title at the same Olympic Stadium where Jesse Owens won four gold medals at the 1936 Berlin Olympic Games.

Several researchers have done studies to predict how fast a man can run 100 meters. Most models are based on the idea that we are getting close to the limit because the amount of improvement is decreasing. Bolt's results changed this perception.

Bolt, who has set three records at the 100 meter distance with times of 9.72, 9.69, and 9.58, is already looking to go far below some estimates. He thinks the world records will stop at 9.40. Some bookmakers are betting that Bolt will get there. The method for estimating the ultimate record presented in this article gives a 90% confidence interval for the ultimate record that has a lower bound of 9.40.

To measure the effect Bolt has had so far and may have in future we calculated the lower bound excluding his three records. The result is 9.62, greater than the record he set, so Bolt has already beaten the estimated ultimate record based on other runner's times. We have also carried out some other probability calculations by excluding his records. They show how remarkable Bolt's performance has been. He has become the premier runner in the world and has changed our perception of human capabilities.

In the following sections we will present methods we consider relevant for forecasting future records. Other methods can be found in [?, 6, 9, 10] and references therein.

23.2   Methods Based on Trend Analysis

Sports records have improved over the years, often faster than our expectation. To analyze the data many investigators have used models made up of a deterministic component, to account for the trend, and a stochastic component, to account for the random variation. Some attempts have also been made to estimate the limiting times (ultimate records) using models with exponential-decay deterministic component. Although useful it is demonstrated that trend analysis would not produce meaningful performance estimates for future records.

23.3   Methods Based on Outstanding Values

Outstanding sports achievements are usually analyzed using one of the following three main methods:

1. The *Extreme Value Theory* that usually deals with maxima or minima. This method uses the absolute largest or absolute smallest values in a specific time period.

2. The *Threshold Theory* that deals with values above or below a specified threshold.

3. The *Theory of Records* that deals with values larger or smaller than all the previous values.

The frequency of outstanding values is analyzed using the *Theory of Exceedances*. This theory deals exclusively with the number of times a chosen threshold is exceeded.

23.3.1 Methods Based on Threshold Theory

In this approach the probabilities of future performances are calculated by developing models for a tail of the distribution for performance measures. Since performance measures above or below a threshold carry more information about future exceptional performances, methods based on tail modeling are appropriate. Methods based on this approach assume that the tail of the distribution for the performance measure of interest belongs to a parametric family and carry out the inference using excesses, that is, the performance measures greater or smaller than some predetermined value y_0. It has been shown that the natural parametric family of distributions to consider for excesses is the generalized Pareto distribution (GPD);

$$P(Y \leq y) = 1 - \left(1 - \frac{ky}{\sigma}\right)^{1/k}$$

where Y represents a performance measure and $\sigma > 0$ and $-\infty < k < \infty$ are unknown parameters [7]. The range of Y is $0 < y < \infty$ for $k \leq 0$, and $0 < y < \sigma/k$ for $k > 0$.

For example, for the men's 100 meter dash a threshold such as 10 seconds may be considered. The GPD includes three specific forms:

1. Long tail Pareto distribution.

2. Medium tail exponential distribution.

3. Short tail distribution with an endpoint.

Most classical distributions have tails that behave like one of these forms. Unfortunately, like most asymptotic results, applying this approach is not free of problems. The obvious problems are the choice of a parametric family, determination of the threshold, and having to deal with intractable equations that need to be solved to obtain estimates of the parameters.

Hill [5] and Davis and Resnick [2] have proposed an approach that is easy to use and is applicable to a wide class of distribution functions possessing medium or long tails. They propose assuming a tail model of the form $F(y) = cy^{-a}$, $y > y_0$. It is a suitable model for men's 100 meter data considering that the recent records set by Bolt indicate a long tail.

From a random sample $Y_1, Y_2, \ldots, Y_n$ the estimates of the parameters are obtained based on the upper $m = m(n)$ order statistics (best times) where m is a sequence of integers chosen such that $m \to \infty$ and $m/n \to 0$. Here c is estimated using the empirical $1m/n$ quantile, $Y_{(m+1)}$ and $1/a$ is estimated by

$$\hat{a} * (n/m) = m^{-1} \sum_{i=1}^{m} \ln Y_{(i)} - \ln Y_{(m+1)}.$$

Statistical theory regarding these estimators is well established. The Pareto-tail upper tail estimate $\bar{F}(y)$ is then

$$\bar{F}(y) = \frac{m}{n} \left(\frac{y}{Y_{(m+1)}} \right)^{-1/\hat{a}*(n/m)} \qquad Y_{(m+1)}.$$

The lower tail estimate can be obtained similarly.

Application to the Men's 100 Meter Dash   Consider the data for the men's 100 meter after January 1, 1977, when the International Association of Athletics Federations (IAAF) required fully automatic timing to the hundredth of a second. From January 1, 1977 to September 1, 2009 there are between 30 to 40 data points, some being invalidated due to drug use and some being discarded (including Gay's 9.68 time on June 2008 during the 2008 U.S. Olympic Trials) due to wind speeds exceeding the IAAF legal limit. Here the time between the two latest records set by Bolt and the time his recent record has stood to date are inordinately small. So rather than using an $m(n)$ that involves these values here we use $\sqrt{n}$ instead. This is a good choice as the rate of convergence is the same for both requirements. For the data considered, the closest integer to $\sqrt{n}$ is 6 ($m(n) = 6$). The top six legal times are

$$Y_1 = 9.58, \; y_2 = 9.69, \; y_3 = 9.71, \; y_4 = 9.72, \; y_5 = 9.72, \; y_6 = 9.74$$

set respectively by Bolt, Bolt, Gay, Powell, Bolt, and Powell. Using these for $y < y_7 = 9.77$ we get the tail model

$$\bar{F}(y) = 6/33(9.77/y)^{-126.694}.$$

Using this $P(Y \leq 9.5)$ and $P(Y \leq 9.55)$ are respectively 0.0052 and 0.0102.

Bolt's Effect To measure the effect Bolt has had so far and may have in the future let us calculate the same probabilities excluding his three records. The top six legal times are then

$$y_1 = 9.71, \ y_2 = 9.72, \ y_3 = 9.74, \ y_4 = 9.77, \ y_5 = 9.79, \ y_6 = 9.84.$$

set respectively by Gay, Powell, Powell, Powell, Greene, and Bailey. Using these $y < y_7 = 9.84$ we get the tail model

$$\bar{F}(y) = 6/33(9.84/y)^{-124.954}.$$

The values of $P(X \leq 9.5)$ and $P(X \leq 9.55)$ are now respectively 0.00225 and 0.00433. These are significantly lower than the probabilities we get using Bolt's records. Also we have $P(X \leq 9.58) = 0.0064$, which demonstrates that Bolt is in a different league.

23.3.2 Methods Based on Theory of Records

The theory of records deals with values that are strictly greater than or less than all previous values. Usually the first value is counted as a record. Then a value is a record (upper record or record high) if it exceeds all previous values.

The study of record values, their frequencies, times of their occurrences, their distances from each other, etc. constitutes the theory of records. Formally the theory deals with four main random variables:

1. The number of records in a sequence of n observations.

2. The record times (indices).

3. The waiting time between the records.

4. The record values.

The theory of records has not yet been fully exploited to address questions regarding the prediction of sports records because the results of the theory of records for independent and identically distributed sequences are not directly applicable to sports. In fact, in most sports records occur more frequently than what the theory predicts. To account for this one may treat the problem as if participation has increased with time, or more attempts are made so that the probability of setting a new record was increased. [6]

To predict future records, Noubary [6] has developed a simple method using the following results of the theory of records for an independent and identically distributed sequence of observations:

(a) If there is an initial sequence of n_1 observations and a batch of n_2 future observations, then the probability that the additional batch contains a new record is $n_2/(n_1 + n_2)$.

(b) For large n, $P_{r,n}$, the probability that a series of length n contains exactly r records is given by

$$P_{r,n} \sim \frac{1}{(r-1)!n} (\ln(n) + \gamma)^{r-1}$$

Where $\gamma = 0.5772$ is Euler's constant. To illustrate the method, consider the 100 meter data for the period 1912–2009. The number of records set is $r = 20$. The maximum likelihood method yields $n = 317,884,920$ attempts. This is found by maximizing $P_{r,n}$ with respect to n. It estimates the number of independent and identically distributed attempts required to produce twenty records.

Now suppose, for example that, i is the geometric rate of increase per year due to increase in the number of attempts. This means that the number of attempts in any given year is assumed to be i times the number of attempts in the year before. The value of i can be found by solving the equation

$$1 + i + i^2 + \cdots + i^{97} = 317884920,$$

where i^j represents the number of attempts in year j. We get $i = 1.2013$, which means 20.13% more attempts per year. To predict records for the future one and five years (in this case year 2010 and the period 2010–2015) we replace $n_1 = 317884920$ and $n_2 = (1.2013)^{98} = 63950619$ for the next year, and $n_1 = 317884920$ and $n_2 = (1.2013)^{98} + \cdots + (1.2013)^{102} = 477112800$ for the next five years in (a). This leads to probability estimates of 0.1926 and 0.6001 for a new record in the next one and five years, respectively.

23.4 Ultimate Record

Prediction of ultimate records can be carried out using a model with an exponential-decay trend. It is also possible to consider simpler trends such as

$$y = (b + ct)/(1 + t) \quad \text{or} \quad y = (a + bt + ct^2)/(1 + t + t^2)$$

and use the fact that as $t \to \infty$, $y \to c$. However, the application of such models usually results in predictions with large standard errors that are not useful or even acceptable [8]. This section describes a method based on tail modeling using a certain number of exceptional performances. This avoids the above mentioned problem and provides a confidence interval for the ultimate record based on the most recent performances or records.

Let Y and $Y_1, Y_2, \ldots, Y_n$ represent, respectively, a performance measure and a set of observed values for a given sport such as the 100 meter dash where

$$Y_1 \leq Y_2 \leq \cdots \leq Y_n.$$

Assuming that the distribution function $F(y)$ has a lower endpoint and certain conditions are satisfied, a level $(1p)$ confidence interval for the minimum of Y is (De Haan [3])

$$\{Y_1 - (Y_2 - Y_1)/[(1-p)^{-k} - 1], Y_1\}.$$

De Haan [3] has also shown that

$$\frac{\ln m(n)}{\ln\left[(y_{m(n)} - y_3)/(y_3 - y_2)\right]}$$

is a good estimate for k.

For example, suppose that we wish to estimate the ultimate record for the 100 meter dash. We consider the same data as in Section 23.3.1. We need to choose an integer $m(n)$ depending on n such that $m(n) \to \infty$ and $m(n)/n \to 0$ as $n \to \infty$. Again we choose $\sqrt{n}$ as it has the same rate of convergence for both requirements. For the data used the closest integer to $\sqrt{n}$ is 6 $(m(n) = 6)$ and the top six legal times are:

$$y_1 = 9.58, \ y_2 = 9.69, \ y_3 = 9.71, \ y_4 = 9.72, \ y_5 = 9.72, \ y_6 = 9.74.$$

Using the above formula we get $k = 4.419$ and a 90% one-sided prediction interval for the ultimate record is (9.40, 9.58).

As mentioned earlier Bolt believes that the world record will not go below 9.40 and the British bookmakers are betting that Bolt will get there.

Bolt's Effect   Now to measure the effect Bolt has had so far and may have in future let us calculate the prediction interval by excluding his three records. The top six legal times are then

$$y_1 = 9.71, \ y_2 = 9.72, \ y_3 = 9.74, \ y_4 = 9.77, \ y_5 = 9.79, \ y_6 = 9.84.$$

set respectively by Gay, Powell, Powell, Powell, Greene, and Bailey. Using the formula we get $k = 1.113$ and a 90% one-sided prediction interval for the ultimate record is $(9.62, 9.71)$. The lower bound is greater than the present record set by Bolt, so he has already exceeded what would have been predicted to be the limit of human performance. This shows how remarkable Bolt's performance has been and how this one runner can changeour perception of human capabilities.

Acknowledgments   I would like to thank Professor Joe Gallian for providing me with help and direction.

References

[1] S.M. Berry, A statistician reads the sports pages, *Chance* 15 (2002) 49–53.

[2] R. Davis, and S. Resnick, Tail estimates motivated by extreme-value theory. *Annals of Statistics* 12 (1984) 1467–1487.

[3] L. DeHaan, Estimation of the minimum of a function using order statistics. *Journal of the American Statistical Association* 76 (1981) 467–469.

[4] J.H.J. Einmahl, and J.R. Magnus, Records in athletics through extreme-value theory, *Journal of the American Statistical Association* 103 (2008) 484–494.

[5] B. M. Hill, A Simple general approach to inference about the tail of a distribution, *Annals of Statistics* 3 (1975) 1163–1174.

[6] R. Noubary, A procedure for prediction of sports records, *J. of Quantitative Analysis in Sports* 1 (2005) 1–12.

[7] J. Pickands, Statistical inference using extreme order statistics, *Annals of Statistics* 3 (1975) 131–199.

[8] R. L. Smith, Forecasting records by maximum likelihood, *J. of the Amer. Stat. Assoc.* 83 (1988) 331–338.

[9] A. R. Solow and W. Smith, How surprising is a new record, *The Amer. Statistician* 59 (2005) 153–155.

[10] J. T. Terpstra and N. D. Schauer, A simple random walk model for predicting track and field World records, *J. of Quantitative Analysis in Sports* 3 (2007) 1–16.

About the Author

Reza D. Noubary received his B.Sc. and M.Sc. in Mathematics from Tehran University, and M.Sc. and Ph.D. in statistics from Manchester University. His research interests include time series analysis, modeling and risk analysis of natural disasters, and applications of mathematics and statistics in sports. He is a fellow of the Alexander von Humboldt Foundation. He frequently teaches courses based on sports. His outside interests include soccer, racquetball, and tennis.

CHAPTER 24

May the Best Team Win: Determining the Winner of a Cross Country Race

Stephen Szydlik

Abstract

Finding the winner of an athletic contest or sporting event should be a simple process: the competitor with the most runs, goals, points, or the quickest time should be the winner. In practice, however, choosing a team winner can be a challenge, especially when many teams are involved in a tournament. In this article, we explore some of the specific dilemmas associated with finding the winning team in a cross country running race.

Standard scoring of a cross country race is straightforward. A team typically consists of seven runners, and a team's score is the sum of the placings of its first five runners. A team's sixth and seventh runners do not score points towards their team's total, but their place can serve to increase the team score of their opponents. Teams are ranked by the order of their scores from lowest to highest. While teams often face off head-to-head in dual meets, invitationals of up to 30 teams are also common.

Though simple to implement, this race scoring system can yield surprising and somewhat counterintuitive outcomes. Chief among these are failures of *binary independence*: the relative ranking of two teams in an invitational can depend upon the presence and performance of the other teams in the race. For example, one team (the Acorns) might finish ahead of a second team (the Buckeyes) when a third team (the Chestnuts) has a good day. However, with identical races by the Acorns and Buckeyes, it is possible for the Acorns to finish behind the Buckeyes if the Chestnuts instead have a bad day. Troubling situations such as these are relatively common in cross country running, and are well known to coaches and interested observers. In this article, we study these issues and consider some alternatives to the standard race scoring methods.

Social choice theory, an area of study on the interface of mathematics, political science, and economics, provides a useful lens through which we can examine cross-country race scoring. One particular aspect of this discipline is concerned with studying the science of decision-making in voting and elections, and comparisons with cross-country scoring are striking. It is arguable, for example, that had Ralph Nader not been a candidate for the U.S. predisdency in the 2000 election, then Al Gore would have beaten George W. Bush. Our electoral system violates binary independence in precisely the way that cross-country scoring does. We examine parallels between cross-country scoring and voting methods with an eye towards Kenneth Arrow's seminal result on the impossibility of finding a completely fair democratic election method. We also draw some important distinctions between the two subjects, particularly in the case of a two-team race.

24.1 Warming Up

Collegiate cross country running in the upper Midwest is highly competitive. In the 36 years (through the fall of 2008), that the National Collegiate Athletic Association (NCAA) has sponsored a Division III (non-scholarship) men's team championship, the top team has come from Illinois, Wisconsin, or Michigan 26 times [6]. The University of Wisconsin-Oshkosh had won three championships in the years leading up to the fall of 2001, when this story begins. The Oshkosh team that year had a great mix of talent and experience, and I had hopes that they would contend for the national title. One early-season test for the Titans was at the Jim Drews Invitational, a race bringing together much of the top Midwest Division III talent. The event was hosted by Wisconsin-LaCrosse, a perennial running powerhouse and nemesis of Oshkosh. Also competing was a team from the University of Wisconsin-Madison, a strong Division I school.

The race took place on a wet Saturday in October, with 33 teams and 223 team runners racing over a 5 mile (8 kilometer) course. At the conclusion of the event, the UW-Madison runners had finished in places 1, 2, 3, 8, and 27, those from UW-LaCrosse had finished in positions 4, 12, 15, 24, 35, 49, and 55, and UW-Oshkosh runners had finished in places 10, 11, 13, 28, 30, 43, and 69. Scoring an NCAA cross country meet is straightforward, with three fundamental rules (See [3, p. 122]):

1. A team's score is the sum of the placings of their first five runners.

2. Teams are ranked by the order of their scores from lowest to highest.

3. Although the sixth and seventh runners of a team to finish do not score points toward their team's total, their place, if better than those of any of the first five of an opposing team, serve to increase the team score of the opponents. When it occurs, this situation is called *displacement*.

Using this scoring method, UW-Madison won the invitational with an aggregate score of 41, UW-LaCrosse's total of 90 placed them second, and

UW-Oshkosh was a close third with 92 points. The results were disappointing for me, but only because I had hoped Oshkosh would beat UW-LaCrosse. Beating UW-Madison was not a realistic goal, and was in fact irrelevant, as the teams compete in different divisions and would not race each other again. This led to an interesting question: if we eliminated the UW-Madison runners from the race results, would the result be the same? After such a change, UW-LaCrosse runners move up to respective places 1, 8, 11, 20, 30, 44, and 50, while the UW-Oshkosh runners improve to positions 6, 7, 9, 23, 25, 38, and 64, so the teams tie for first place, with 70 points each. Somehow the presence of UW-Madison runners was detrimental to the UW-Oshkosh cause, not just because UW-Madison beat UW-Oshkosh, but also because they helped UW-LaCrosse relative to UW-Oshkosh.

What happens if we ignore all other teams except for LaCrosse and Oshkosh? If we are really trying to determine which of the two teams was stronger on that October day, it makes sense to consider them head-to-head, in a two-team race. A way to picture this reduced race is as a finite sequence of runners:

$$\begin{pmatrix} 1 & 2 & 3 & 4 & 5 & 6 & 7 & 8 & 9 & 10 & 11 & 12 & 13 & 14 \\ L_1, & O_1, & O_2, & L_2, & O_3, & L_3, & L_4, & O_4, & O_5, & L_5, & O_6, & L_6, & L_7, & O_7 \end{pmatrix}$$

Here O_i and L_i represent the ith finishers for Oshkosh and LaCrosse, respectively, and the numbers o_i and ℓ_i above them represent their respective finishing positions. Scoring this head-to-head race yields 27 points for Oshkosh and 28 points for LaCrosse. If Oshkosh and LaCrosse had been the only two teams on the course that day, Oshkosh would have won!

Races like the Jim Drews Invitational occur in high school and collegiate cross country meets across the country many times every fall weekend, and scoring oddities such as those involving UW-Oshkosh and UW-LaCrosse are well known to coaches and interested observers. In this article, we will analyze some of the mathematics of cross country scoring, consider criteria that measure the reasonableness of scoring methods, and explore alternative scoring methods. Readers familiar with the language and theorems of social choice theory will note strong parallels to that area of mathematics, and we will borrow liberally from its terminology. We will also, however, draw some strong distinctions between cross country scoring and vote counting.

24.2   Mile 1: Basic Terminology and Some Alternatives

At the start of a large invitational race, runners need to balance carefully two goals: they need to begin quickly enough to establish a favorable position,

but not so quickly that they tire and fade at the end of the race. So as the
starting gun goes off, we clarify what we mean by a "scoring method," and
formalize some of our notations.

Definition 1 *Let* $T = \{T_1, T_2, \ldots, T_n\}$ *be a set of* n *teams, each consisting
of* k *runners. A* race *is an ordered list of the* $n \cdot k$ *runners. A* scoring method
*is a function that accepts a race as an input and produces as output a linear
ordering of* T. *We shall use* $T_{i_1} \prec T_{i_2} \prec \cdots \prec T_{i_n}$ *to denote the ordering,
where team* T_{i_1} *finishes first, team* T_{i_2} *finishes second, and so on.*

In this article, we will insist that the ordering be strict. While team ties occur
during meets, they are easy to break. For example, the NCAA tiebreaking
rule is to compare the tying teams' fifth runners and break the tie in favor
of the team whose fifth runner finishes ahead of the other(s). For simplicity,
we will also consider only scoring methods that are symmetric with respect
to the teams, and anonymous with respect to the runners. By *symmetric*, we
mean that the names of the teams are irrelevant, i.e., interchanging the runners of teams A and B in a race will interchange A and B in the team ranking
outcome. By *anonymous*, we mean that if any two runners from a team interchange places, the outcome remains the same. Anonymity guarantees that
there are $(_{nk}C_k)(_{n(k-1)}C_k) \cdots (_kC_k)$ essentially distinct races.

In scoring a meet, the choice by the NCAA to use five scoring runners
and two displacing runners is somewhat arbitrary. The simplest change we
can make to the usual scoring method is to change those numbers: let us call
a method that uses standard scoring but with m scoring runners and ℓ displacing runners an (m, ℓ)-standard scoring system. For example, the recent
world cross country championships in Amman, Jordan in March 2009 used a
$(4, 2)$-scoring system (see [4]) to determine the team champion. The choice
of k and ℓ can have a great impact on deciding the winning team in a race
even if there are only two teams involved (a *dual meet*). In the head-to-head
matchup between UW-Oshkosh and UW-LaCrosse, Oshkosh would win using $(5, 2)$, $(6, 1)$ or $(3, 0)$-standard scoring, whereas UW-LaCrosse would
win (by virtue of the standard tiebreaker) using $(4, 0)$ scoring. UW-LaCrosse
would also win using $(1, 0)$ standard scoring (since LaCrosse's top finisher
beat Oshkosh's top finisher) but that is hardly an acceptable method because
it depends only on the relative finishing positions of one runner from each
team (in this case, the winning team is the team whose first runner finishes
first). The notion of single-runner dominance in a race will be important for
us later, so let us formalize this idea:

Definition 2 *A scoring method is called* non-team *oriented if there is some
i, $1 \leq i \leq k$ such that for every race, the order of the teams is determined*

solely by the relative positions of the ith finisher for the teams. That is, for all teams A, B in an invitational meet, $A \prec B \Leftrightarrow a_i < b_i$.

For example, in a race involving five runners per team where we ignore the placings of the first four runners from each team and rank the teams based solely on the ordering of their fifth runner would be a non-team oriented scoring method. We would like to avoid such methods!

The Jim Drews Invitational example points to difficulties with the standard invitational scoring methods. Most obvious is that one team's performance relative to another team can depend on the other teams in the field (as UW-Oshkosh's performance relative to UW-LaCrosse was influenced by the presence of UW-Madison). We call this a failure of the *Binary Independence* condition.

Binary Independence condition: *The relative ranking of teams A and B in the race results should not depend upon any other team.*

More precisely, suppose a race takes place in which team A beats team B in the final ranking. Shortly thereafter a second race takes place with the same teams and the same runners, and in that race some runners change positions, but the relative positions of runners from team A and B don't change at all. Then team A should still beat team B in the rankings.

A scoring method should ideally possess this property because deciding whether team A or team B finishes higher in the rankings should not depend on factors such as injury, illness, stumble, or poor performance by a member of team C. If the relative ranking of A and B is never affected by the positions of other runners in any race, then the scoring method is said to satisfy the Binary Independence condition. We use the word "should" in its statement (and in our statements of other conditions in this article) to emphasize that, while desirable, the property is not possessed by all scoring methods. The Jim Drews example shows that $(5, 2)$ standard scoring violates the Binary Independence condition, since moving UW-Madison runners to the end of the race changes the relative ranking of UW-LaCrosse and UW-Oshkosh to a tie, broken in favor of UW-Oshkosh. The other (m, ℓ) standard scoring methods are similarly flawed.

The Binary Independence condition suggests an alternative to the standard invitational scoring methods. Given a race with two or more teams, why not match each team head-to-head with every other team, and declare the team that wins all of those matchups as the winner of the invitational? This amounts to scoring a series of dual meets, with the winner of the invitational being the team that wins all its head-to-head matchups. Second place could be given to the team that wins all but one of its head-to-heads, third to the team that beats all but the top two teams head-to-head, and so on.

Though attractive at first glance, this scoring mechanism has a fatal problem. Consider the following race between three teams, A, B, and C:

EXAMPLE 1

$$\left\langle \begin{array}{ccccccccccccccc} 1 & 2 & 3 & 4 & 5 & 6 & 7 & 8 & 9 & 10 & 11 & 12 & 13 & 14 & 15 \\ A_1, & B_1, & C_1, & B_2, & C_2, & A_2, & C_3, & A_3, & B_3, & A_4, & B_4, & C_4, & B_5, & C_5, & A_5 \end{array} \right\rangle$$

If we match each team head-to-head, computation shows that team A beats team B (27-28), team B beats team C (26-29), and team C beats team A (27-28). In this example, we have a nontransitive result: team A beats team B and team B beats team C, but A fails to beat C. There is no team that wins all of its head-to-head matchups. Thus, matching the teams up in this way does not provide us with a scoring method at all, for it does not provide a linear ordering of the teams.

Though matching the teams up head-to-head fails as a scoring method, it suggests another condition for our scoring methods. Though not every invitational has a team that wins all of its head-to-head matchups, if there is such a team, it is reasonable to expect that team to win the invitational. We shall call this condition the *Condorcet Criterion*, and call a team that wins all its head-to-head matchups a *Condorcet team*.

The Condorcet Criterion: *A team that beats all the other teams in an invitational in head-to-head competition should be the winner of the invitational.*

Standard invitational scoring methods do not possess the Condorcet property in general. Creating examples that show violations is a relatively simple matter. Given a (m, ℓ) standard scoring method (with $m \geq 2$), we find a team A that consists primarily of strong runners but also possessing one or two weak runners. The strong runners allow A to win individual head-to-head matchups with other teams, but if we choose a sufficiently large set of teams $\mathcal{T}$ for our invitational, we can ensure that A's bottom runners will finish far enough down in the rankings that A fails to win the invitational. For example, we can show that the standard $(5, 2)$ scoring method violates the Condorcet criterion using an example involving four teams A, B, C, and D.

EXAMPLE 2

$$\left\langle \begin{array}{cccccccccccccc} 1 & 2 & 3 & 4 & 5 & 6 & 7 & 8 & 9 & 10 & 11 & 12 & 13 & 14 \\ A_1, & A_2, & A_3, & B_1, & C_1, & D_1, & C_2, & D_2, & B_2, & D_3, & B_3, & C_3, & C_4, & B_4, \end{array} \right.$$

$$\left. \begin{array}{cccccccccccccc} 15 & 16 & 17 & 18 & 19 & 20 & 21 & 22 & 23 & 24 & 25 & 26 & 27 & 28 \\ D_4, & B_5, & D_5, & B_6, & C_5, & B_7, & C_6, & C_7, & D_6, & D_7, & A_4, & A_5 & A_6 & A_7 \end{array} \right\rangle$$

In this race, runners from team A take the top three places, but the team is fully displaced, that is, the seventh runner from each of the other three teams finishes before A's fourth and fifth runner. In each head-to-head competition involving A then, A takes places 1, 2, 3, 11, and 12 giving it an aggregate score of 29. Each opposing team in the head-to-head with A places runners in positions 4 through 10, giving it a score of 30. Thus A wins each head-to-head matchup and is therefore a Condorcet team. However, when we score the race using the standard method, A has 57 points, B 54 points, and C and D 56. So A not only fails to win the invitational, but actually finishes last. Standard invitational scoring violates the Condorcet criterion, and it can do so in the worst possible way!

24.3   Mile 2: Fairness Criteria and Other Scoring Methods

The first mile of a cross country race typically goes quickly. Runners are fresh and the excitement of competition carries them along. In the second mile, runners need to assess their positions, while remembering to be patient as much of the race remains. In our consideration of scoring methods, we'll consider some other slightly more complex scoring methods and ways to measure their reasonableness. But we'll need to be patient.

Let us begin our second mile by considering a slightly different way of comparing two teams, call them P and Q, head-to-head. One way to measure an individual runner's strength is to count how many opponents he is beaten by, ignoring his own team's runners. A team's score could then be determined by aggregating this information among all the team's scoring runners. More formally, given that runner P_i is the ith scoring runner (of m) from team P, let $p_{i,Q}$ be the number of runners from team Q who finish ahead of runner P_i. Then $0 \leq p_{i,Q} \leq m + \ell$, where ℓ is the number of non-scoring displacing runners. A team score for P in a head-to-head matchup with team Q can be defined as $\sum_{i=1}^{k} p_{i,Q}$, with the lower overall score winning the meet. Let us call this method the (m, ℓ) *Runner Matchup Method*, since each individual runner on a team is compared with the opposing team.

If we apply the $(5, 2)$-Runner Matchup Method in a dual meet between teams A and B from Example 2, we see that $a_{1,B} = 0$, $a_{2,B} = 0$, $a_{3,B} = 0$, $a_{4,B} = 7$, and $a_{5,B} = 7$. Using the Runner Matchup Method, team A scores 14 points. A similar calculation shows that team B scores 15 points.

Is this really a new scoring system? An alert reader will notice that the team tallies in this example (14-15) differ by 15 points from the respective tallies (29-30) obtained using the standard head-to-head scoring. This is no coincidence. Let P and Q be two teams in a dual meet. Given runner P_i from team P ($0 \leq i \leq m$), let p_i denote P_i's overall finishing place. Since P_i's place in a dual meet with Q is determined by the runners finishing ahead of him (both on team P and on team Q), we have $p_i = i + p_{i,Q}$. Thus the (m, ℓ) standard team score for P in a dual meet with Q is

$$\sum_{i=1}^{m} p_i = \sum_{i=1}^{m} (i + p_{i,Q}) = \frac{m(m+1)}{2} + \sum_{i=1}^{m} p_{i,Q}.$$

Thus $\sum_{i=1}^{m} p_i$ and $\sum_{i=1}^{m} p_{i,Q}$ differ by a term that depends only on m, the number of scoring runners, and not on the placements in any individual race. Thus the (m, ℓ) standard scoring method and the (m, ℓ) Runner Matchup Method are equivalent, yielding identical team rankings and identical differences between team scores in dual meets.

Though the Runner Matchup Methods are no different from the standard scoring methods, they allow us to identify a powerful relationship between a team's aggregate score in an invitational and its head-to-head scores against other teams in the race. To see it, suppose that in an invitational $|T| = n$ and each team has m scoring runners. As in the dual meet, a runner's finishing place in an invitational is determined by runners finishing ahead of him, both on his and on opposing teams. Thus,

$$p_i = i + \sum_{\substack{Q \in T \\ Q \neq P}} p_{i,Q}.$$

Then the invitational team score for team P using the (m, ℓ) standard scoring method is

$$\sum_{i=1}^{m} p_i = \sum_{i=1}^{m} \left(i + \sum_{\substack{Q \in T \\ Q \neq P}} p_{i,Q} \right)$$

$$= \sum_{i=1}^{m} i + \sum_{\substack{Q \in T \\ Q \neq P}} \left(\sum_{i=1}^{m} p_{i,Q} \right)$$

$$= \sum_{i=1}^{m} i + \sum_{\substack{Q \in T \\ Q \neq P}} \left[(\text{Standard Team Score of } P \text{ vs. } Q) - \sum_{i=1}^{m} i \right]$$

$$= (|T| - 2) \sum_{i=1}^{m} i + \sum_{\substack{Q \in T \\ Q \neq P}} (\text{Standard Team Score of } P \text{ vs. } Q).$$

These show that teams' standard invitational scores can be obtained from the sum of their standard head-to-head scores. We need only subtract a constant based on the number of teams in the invitational. We thus have

THEOREM 1 *A team's score using standard invitational scoring can be determined from its head-to-head matchup scores. If there are n teams and m scoring runners per team (with or without displacement), then a team's invitational score is the sum of its head-to-head scores, minus* $\frac{(n-2)m(m+1)}{2}$.

EXAMPLE 3 Using the data from Example 2, there are $n = 4$ teams and $m = 5$ scoring runners. Team A scores 29 points in each of its head-to-head matchups with teams B, C, and D. The sum of these scores is 87; from this we subtract $\frac{(n-2)(m)(m+1)}{2} = \frac{(4-2)(5)(5+1)}{2} = 30$ to find A's invitational total of 57. The other team's invitational scores may be similarly verified.

Aside from describing a relationship between a team's head-to-head scores and its aggregate score in an invitational, Theorem 1 yields several corollaries, the first of which is immediate. Recall that a Condorcet team in an invitational beats every other team head-to-head.

COROLLARY 1 *Using (m, ℓ) standard scoring, the winning team in an invitational is the team whose head-to-head scores have minimal sum, or equivalantly, have minimal average.*

COROLLARY 2 *The $(m, 0)$ standard scoring method cannot rank a Condorcet team at the bottom.*

Proof. Without displacement, the total number of points available to both teams in every head-to-head matchup is the same: $\sum_{i=1}^{2m} i = m(2m + 1)$. Call this number S. If team P wins all its head-to-head matchups, then P will score fewer than $\frac{S}{2}$ in each of its matchups, and so its overall average will be less than $\frac{S}{2}$ as well. The average score of all n teams across all matchups must be $\frac{S}{2}$, so there must be another team, call it Q, whose average is greater than S. Thus team Q will be ranked below P in the invitational rankings.

Corollary 2 is important because it demonstrates a problem with using displacements. In Example 2, team A was bottom ranked even though it won all of its head-to-head matchups. Corollary 2 tells us that this is a consequence of allowing displacing runners. Though the $(m, 0)$ standard scoring methods fail the Condorcet criterion in general, at least they cannot fail as badly as those (m, ℓ) methods where $\ell > 0$: A Condorcet team cannot be bottom ranked.

24.4   Mile 3: More Criteria and Alternative Scoring Methods

Since a Condorcet team A in an invitational beats every other team head-to-head, it is natural to ask if there are any methods that guarantee team A a first place finish. Consider the following scoring procedure: score the teams using a standard $(m, 0)$ scoring method. Once a preliminary ranking of the teams has been established, we drop the lowest ranked team. Since A cannot be ranked last in this method, A will not be eliminated. Now we rescore, with all of the runners from the bottom-ranked team removed. A remains the Condorcet team after this elimination step. We now repeat the process, scoring the invitational as usual and eliminating the team that is now bottom-ranked. We continue until there is only one team remaining, and declare that team to be the winner of the invitational. In this case, we know that the winning team must be A: at every stage, team A remains the Condorcet team, and by Corollary 2 cannot be bottom-ranked at any step in the process. Thus team A can never be eliminated. If we call this method the *Iterated $(m, 0)$ Scoring Method*, then we have shown the following result.

COROLLARY 3 *The iterated $(m, 0)$ scoring methods satisfy the Condorcet criterion.*

We can extend this to obtain a complete ranking of the teams in the invitational. Once the top-ranked team has been determined, we can eliminate that team and determine the iterated $(m, 0)$ scoring method winner from among the remaining teams, calling that the second place team. This ensures that a team that beats all other teams head-to-head except A is guaranteed to finish second, if such a team exists. Other rankings may be determined similarly.

EXAMPLE 4 Consider the following race:

$$\begin{pmatrix} 1 & 2 & 3 & 4 & 5 & 6 & 7 & 8 & 9 & 10 & 11 & 12 & 13 & 14 & 15 \\ A_1, & A_2, & A_3, & B_1, & B_2, & B_3, & B_4, & B_5, & C_1, & C_2, & C_3, & C_4, & C_5, & A_4, & A_5 \end{pmatrix}$$

In head-to-head competition, A beats teams B and C by identical 25-30 scores and is therefore the Condorcet team. The team ranking using $(5, 0)$-standard scoring is $B \prec A \prec C$ by a score of 30-35-55. This shows that we may have a violation of the Condorcet criterion even if displacement is not allowed. However, if we use the iterated $(5, 0)$ scoring method, then team C is dropped at the end of the first round, and teams A and B face off in the second round with A winning. To find the second place team, we allow B and C to face off head-to-head (with A removed). Then B beats C 15-40, so the ranking using the iterated (5,0) scoring method is $A \prec B \prec C$.

We have identified a scoring method that satisfies the Condorcet criterion. The iterated methods are less intuitive than standard scoring (can you imagine coaches trying to predict their teams' places in the middle of a race?), but are significantly more stable. Barring practical concerns, is there any theoretical reason to avoid such a method? Unfortunately, the answer to that question is, "Yes."

EXAMPLE 5 Consider the following race among three closely matched teams A, B, and C. Midway through the race, the placings of the runners are as follows:

$$\begin{pmatrix} 1 & 2 & 3 & 4 & 5 & 6 & 7 & 8 & 9 & 10 & 11 & 12 & 13 & 14 & 15 \\ C_1, & C_2, & A_1, & A_2, & A_3, & B_1, & B_2, & B_3, & B_4, & B_5, & A_4, & C_3, & C_4, & C_5, & A_5 \end{pmatrix}$$

The quick-thinking coach from team A scores the race at the midway point, using the iterated (5,0) scoring method. When she does so, she finds that in the first round of scoring, the scores for A, B, and C are, respectively, 38, 40 and 42. In this preliminary scoring, Team C would be eliminated, and the race would be rescored with only teams A and B. In the second round, A would have runners in positions 1, 2, 3, 9, and 10, for a total score of 25, while B would hold positions 4 through 8 for a total score of 30. So at this point in the race, A is leading the race, but it is close! The coach from team A exhorts her team to push hard over the last half of the course, and one of them, runner A_4, responds. Over the last mile, she surges ahead of all the runners from team B, clinching her team's victory, right? Let's look at the final results.

The final runner placings in the race are

$$\begin{pmatrix} 1 & 2 & 3 & 4 & 5 & 6 & 7 & 8 & 9 & 10 & 11 & 12 & 13 & 14 & 15 \\ C_1, & C_2, & A_1, & A_2, & A_3, & A_4, & B_1, & B_2, & B_3, & B_4, & B_5, & C_3, & C_4, & C_5, & A_5 \end{pmatrix}$$

Using the iterated $(5,0)$ scoring method, Round 1 yields scores of 33, 45, and 42 for teams A, B, and C, respectively. Team B is eliminated and we rescore the race. In Round 2, A holds positions 3, 4, 5, 6, and 10, while C holds places 1, 2, 7, 8, and 9. In the final scoring, then, A has 28 points, while C has 27. Amazingly, C wins! Runner A_4's surge through the field of runners perversely hurt her team's cause.

Unfortunately, situations such as this, though relatively rare, can occur with the iterated scoring methods. We say that they violate the *Monotonicity* criterion.

Monotonicity Criterion: *If team P is the winner of a race, and in a second race with the same teams and runners, a runner from team P improves his performance by moving up one or more places and no other runners change position, then team P should remain the winning team.*

Our search for an improved scoring method is beginning to get frustrating! We have seen flaws in the usual system, but each new method possesses some undesirable properties of its own. Before we get too depressed, however, let us make note of some positives for our scoring systems.

First, the (m, ℓ) standard scoring methods satisfy the monotonicity criterion, for improving a winning team's individual placings can only lower that team's score in these point-based systems, and can only raise other teams' scores. Second, all the scoring methods we have described thus far possess a desirable property that we call the *Pareto* condition.

The Pareto Condition: *If a scoring method involves k runners from each of teams A and B, and if $a_i < b_i$ (that is, if runner A_i beats runner B_i) for all i $(1 \leq i \leq k)$, then team B should not finish above team A in the rankings.*

The Pareto condition says that if all of my team's runners beat their individual counterparts from your team, then your team should not be ranked above my team by the scoring method.

Since better placings in a race yield lower team scores in the standard scoring methods, the (m, ℓ) standard scoring methods, and by extension, the iterated (m, ℓ) scoring methods satisfy the Pareto condition. Surprisingly, however, there are some seemingly reasonable methods that violate it. Consider the following method, another way to match up the teams head-to-head: create some fixed ordering of the teams in the invitational (perhaps randomly). After choosing a (m, ℓ) standard scoring method, match up the first two teams in the list head-to-head using that method. The winner of that matchup will advance to face the third team in the list in another head-to-head matchup. The winner of that will progress to face the fourth team in the list, and so on. The winner is the team that is left when the list has been exhausted.

Not surprisingly, this method satisfies the Condorcet criterion, since any team capable of winning each of its head-to-head matchups will survive to the end of the competition. On the other hand, this sequential-type scoring method violates the Pareto condition in general.

EXAMPLE 6 Consider the following race with four teams (A, B, C, and D) with three runners per team:

$$\left\langle \begin{array}{cccccccccccc} 1 & 2 & 3 & 4 & 5 & 6 & 7 & 8 & 9 & 10 & 11 & 12 \\ C_1, & B_1, & A_1, & B_2, & A_2, & A_3, & D_1, & D_2, & D_3, & C_2, & C_3, & B_3 \end{array} \right\rangle$$

Let us score this meet using $(3, 0)$ standard scoring using head-to-head matchups in the sequence $[A, B, C, D]$. We match up teams A and B (ignoring all the other teams) and find that team B wins, 10-11. Team B advances to face team C which beats B by a head-to-head score of 10-11. Team C then advances to face team D, and D wins that matchup by the score of 9-12. So D is the winner of the invitational using this method.

However, runner A_1 beats runner D_1, A_2 beats D_2, and A_3 beats D_3. This example therefore shows a violation of the Pareto condition. In fact, the offense is far worse than a simple Pareto violation. Using this process, Team D wins the invitational even though *all of team A's runners finish before all of team D's runners*. Similar violations of the Pareto condition using general (m, ℓ) sequential scoring of this type can easily be constructed for $m > 3$ and $\ell > 0$.

24.5 Mile 4: Some Social Choice Theory

The middle miles of a cross country race are often where the race is won. Runners need to keep their attention on their competition even as they become fatigued. Here, we focus on some of the formal language of voting theory. We will use the notation of [9]: [8] provides an elementary introduction to the subject.

We consider the situation in which a group of people (the voters) wishes to decide among several alternatives (most typically, these alternatives are candidates). Let A be the (finite) set of candidates or *alternatives*, and let V be the (finite) set of voters. We assume that each voter in V has an established preference between any two candidates $a, b \in A$, and that these preferences are transitive. Thus the voter has a linearly ordered *preference list* of candidates, ranking them from top to bottom.

Definition 3 *A* social welfare function *is a function that accepts as input a sequence of individual preference lists of some fixed set A and produces as output a single listing (perhaps with ties) of the set A. This list is called the* social preference list.

Social welfare functions provide the means of counting the ballots in an election, and different social welfare functions yield different election results in general. For example, the well-known *Plurality Method* is the social welfare function that ranks the candidates in order of the number of times they appear at the top of voters' preference lists. The *Borda Count* awards points to the candidates as follows: if there are n candidates, the candidate at the bottom of a list gets zero points, the alternative at the next to the bottom spot gets one point, then next one up gets two points, and so on up to the top candidate on the list who gets $n - 1$ points. For each candidate, we add up the total number of points awarded over all the preference lists, and rank the candidates in order of total points, from most to least.

A connection between social welfare functions and cross country scoring methods is immediately apparent. The teams in an invitational can be thought of as the alternatives in an election, with the runners serving as the voters: the number of voters corresponds to the number of runners per team. Finding the winner of an invitational is then analogous to selecting the winner of an election, and cross country scoring methods correspond to social welfare functions. As a simple example, consider a two voter, three candidate election, where voter 1's preferences are candidates B, A, C in that order, while voter 2's preferences are candidates A, B, C. By analogy, we picture this as corresponding to a three-team race with two runners per team: $< B_1, A_1, C_1, A_2, B_2, C_2 >$. We have no need here to make the analogy more precise; rather, we ask the reader only to consider that there is a parallelism between the two domains. In some sense, we can think of the runners in a race as voting with their feet!

While cross country scoring and vote counting share some similarities, they are clearly not the identical. A notable distinction occurs when there are two candidates/teams. May proved (see [2]) that with two candidates, there is only one reasonable method of choosing the winner of an election: majority rule, i.e., the winner of a 2-candidate election should be the candidate who receives more first place votes. However, as we noted in our UW-Oshkosh vs. UW-LaCrosse head-to-head example when we considered different values of

m and ℓ in the (m, ℓ) standard scoring method, there are many reasonable ways to score a two-team race. Moreover, even when the number of scoring runners remains fixed, we can see different results.

EXAMPLE 7 When UW-Oshkosh and UW-LaCrosse are considered head-to-head in the Jim Drews example, Oshkosh runners finish in positions 2, 3, 5, 8, and 9, while LaCrosse runners finish at 1, 4, 6, 7, 10. Rather than finding the team scores using the sum of the placings (in which case Oshkosh wins), we could find the sum of the cube roots of the runners' placings, again giving victory to the team with the smaller score (This effectively places less importance on earlier finishers). LaCrosse wins the head-to-head competition by an approximate score of 8.47 to 8.49.

In this sense, then, there are more cross country scoring systems than there are social welfare functions.

Just as there are criteria measuring the reasonableness of cross country scoring methods, there are means of assessing the value or fairness of different social welfare functions. In fact, our cross country scoring criteria have been drawn directly from analogous criteria in voting theory. For example, the *Pareto Condition* (social choice version) asserts that if every voter prefers candidate x over candidate y on his or her ballot, then y should not win the election. Similarly, a social welfare function is said to satisfy the *Binary Independence Condition* (social choice version) if the following holds: Given our fixed sets A and V, but two different sequences of individual preference lists. Suppose that exactly the same people have candidate x over candidate y in their list. Then we either have x over y in both output rankings of the candidates, or y over x in both output rankings, or y and x tied in both output rankings. That is, the positioning of candidates other than x and y in the individual preference lists is irrelevant to the question of whether x is preferred to y or not in the final output list.

Just as our standard cross country scoring methods' violations of binary independence can have consequences for invitationals, violations of binary independence in the social choice domain can have profound effects on elections. For example, had Ralph Nader not been a U.S. presidential candidate in the 2000 election, then it is likely that Al Gore would have beaten George W. Bush ([7]). The U.S. presidential electoral system violates binary independence in precisely the way that cross-country scoring does.

While fairness criteria and the search for an optimal social welfare function have been the subject of research for well over half a century, Kenneth Arrow proved ([1]) that this exploration is inherently limited.

ARROW'S IMPOSSIBILITY THEOREM (1950)[1] *If A has at least three elements and the set V of individuals is finite, then the only social welfare function for A and V satisfying the Pareto and Binary Independence conditions is one for which there is a single voter $v \in V$ such that for every choice of individual preference lists by the voters, the social preference list is precisely the same as the individual preference list of v.*

Arrow's Theorem says, simply, that the only social welfare functions satisfying both the Pareto and Binary Independence conditions are dictatorships (with v serving as the dictator).

24.6   Mile 5: Impossibility?

The disheartening implication of Arrow's Theorem is that there is no democratic voting method that satisfies both the Pareto and Binary Independence criteria. When we return to the analogy between cross country scoring and vote counting, we naturally wonder whether Arrow's Theorem yields an analogous result in running. We take up this question in the last mile of our race.

In pursuing alternatives to our standard scoring methods, we have encountered numerous roadblocks. Each of our methods suffered from some notable flaw, whether it be a violation of the Condorcet, Monotonicity, or Pareto criteria. Moreover, we have not yet found a cross country scoring method that does satisfy our condition of Binary Independence, the issue that motivated our initial investigation. Before we show how difficult this condition is to satisfy, let us look at an example.

EXAMPLE 8 Suppose that teams A and B are the only two teams in a race, and we allow only $k = 2$ runners per team. There are $_4C_2 = 6$ possible different races:

1. $\langle A_1, A_2, B_1, B_2 \rangle$,   and symmetrically   4. $\langle B_1, B_2, A_1, A_2 \rangle$
2. $\langle A_1, B_1, A_2, B_2 \rangle$,   5. $\langle B_1, A_1, B_2, A_2 \rangle$
3. $\langle A_1, B_1, B_2, A_2 \rangle$,   6. $\langle B_1, A_1, A_2, B_2 \rangle$.

Without knowing precisely what our scoring method is, we can say a number of things about it. For example, if we insist that the Pareto condition be satisfied, team A necessarily must win Races 1 and 2, while team B wins Races 4 and 5. Only Race 3 and Race 6 offer any possible ambiguity. Suppose that our scoring method allows A to win the third race. Then, by symmetry, B

---

[1] Arrow's Impossibility Theorem earned him the Nobel prize in Economics in 1972.

wins Race 6. Recall our definition of *non-team oriented methods* from Definition 2. In this case our scoring method is indeed non-team oriented: in each race, the team winner corresponds to the team whose first runner finishes first. On the other hand, suppose that we decide instead that team A should win race 3 (and, symmetrically, B wins Race 6). This again yields a non-team oriented scoring system, in which the winning team is the team whose *second* runner finishes first, i.e., $A \prec B \Leftrightarrow a_2 < b_2$.

Example 8 shows that when there are two teams and only two runners per team, there are *only* two scoring methods that satisfy the Pareto condition, and both of these are non-team oriented. Though this is disturbing, it is but a single, small example. What if there are more than two teams?

EXAMPLE 9 Suppose that as in Example 8 there are only $k = 2$ runners per team, but now there are three teams in the race. Then there are a total of $(_6C_2) \cdot (_4C_2) = 90$ different races. Consider the race

$$\langle A_1, B_1, B_2, A_2, C_1, C_2 \rangle . \tag{24.1}$$

If we insist that our scoring method satisfy the Pareto condition, then team C should finish last. However, there is flexibility in the relative positions of teams A and B. Suppose that our scoring method chooses A to be the winning team in the race. If our method satisfies the Binary Independence condition (as well as symmetry), then the method is completely determined. For example, consider the race $\langle B_1, A_1, C_1, C_2, B_2, A_2 \rangle$. The Pareto condition ensures that $B \prec A$, while Binary Independence ensures that $B \prec C$, since ignoring team A gives the race $\langle B_1, C_1, C_2, B_2 \rangle$, which B wins by symmetry. Similarly, $A \prec C$, so $B \prec A \prec C$. The team ranking corresponds precisely to the finishing order of the teams' first runners. This is no coincidence: the team rankings from each of the 90 races is determined solely by the first runner placings. So our choice to make A the winner in (24.1) gave us a non-team oriented method depending only on a team's first runners. Similarly, if we had chosen B to be the winner in (24.1), we would have found our method to be non-team oriented, with team rankings dependent only on the teams' second runners.

Examples 8 and 9 highlight the challenge of satisfying both the Pareto and Binary Independence conditions. Unfortunately for seekers of fair scoring methods, Example 9 generalizes to an arbitrary number of scoring runners. This is our main result.

THEOREM 2 **Arrow's Theorem for Cross Country Scoring:** *If there are at least three teams involved in an invitational, then the only scoring methods that satisfy Binary Independence and the Pareto condition are the non-team oriented methods.*

That is, the only way to score a race consistently that will ensure that the Pareto and Binary Independence conditions are satisfied is to pick an index i ($1 \leq i \leq k$), and rank the teams according to the placings of their ith runners. Thus, the social choice version of Arrow's Theorem yields an analogous version in the cross country running domain, with the collection of ith runners for each team playing the role of the dictator in an election. Our discussion in the previous section highlighted some distinctions between elections and cross country races, so the classes of social welfare functions and cross country scoring methods are different sets. Thus Theorem 2 is not merely a restatement of of Arrow's Theorem in a new domain; it is in fact a slightly different result.

The proof of Theorem 2, like the proof of the social choice version of Arrow's Theorem, is not conceptually difficult, but it is technical, so it is omitted here. The result follows using a modification of Taylor's proof of Arrow's Theorem (social choice version), found in §10.4 of [9].

An interesting coincidence is that the traditional symbol of cross country running is the two letters "CC" cut by an arrow (see Figure 24.1). Perhaps the symbol designers knew something about scoring methods! And with that final push, we cross the finish line.

Figure 24.1. The traditional symbol for cross country running

24.7   Warmdown: Some Concluding Remarks

The cross country version of Arrow's Theorem is disheartening. There is no reasonable scoring method that satisfies both the Pareto condition and Binary Independence. All is not lost, however. Here is one very simple way of finding the winner of an invitational: rank the teams in order of *total finish time* or equivalently, average finish time of the scoring runners. This is the system that is used to rank the teams in the International Association of Ultrarunners 100 Kilometer World Challenge events (see §10.24 of [5]). This system satisfies all of our stated reasonableness criteria. However, there is a tradeoff: according to our definition, it is not a scoring method. Scoring methods accept as input a ranking of runners and do not take into account finishing times.

This distinction is as much philosophical as technical. Since this *Total Time Method* involves only the runners' times and not their placings, Binary Independence is satisfied. UW Oshkosh's relative placing against UW LaCrosse does not rely at all upon the presence of UW-Madison runners. However, this method also removes one exciting aspect of cross country running: the head-to-head individual competition. The total time method involves a race against a clock rather than against opponents. So though it might be a mathematically superior means of scoring a race, it is perhaps less desirable.

After competing strongly all season and climbing to second in the final NCAA Division III cross country poll, the UW-Oshkosh men's cross country team finished sixth at the 2001 NCAA Championships in Rock Island, Illinois. The winner? UW-LaCrosse.

Acknowledgments: The author wishes to thank Joe Gallian for his helpful comments in the preparation of this manuscript and Underwood Dudley and John Beam for their careful editing.

References

[1] Kenneth Arrow, A difficulty in the concept of social welfare, *Journal of Political Economy*, 58 (1950) 328–346.

[2] Kenneth May, A set of independent, necessary and sufficient conditions for simple majority decision, *Econometrica*, 20 (1952) 680–684.

[3] The National Collegiate Athletic Association (NCAA), 2009/2010 NCAA men's and women's track and field and cross country rules, 2009 (accessed June 10, 2010). Available at
www.ncaapublications.com/productdownloads/TF10.pdf

[4] The International Association of Athletics Federations (IAAF), 37th IAAF World Cross Country Championships, 2009 (accessed August 27, 2009). Available at www.iaaf.org/WXC09/index.html

[5] International Association of Ultrarunners, Technical guidelines for the organisation of a major IAU competition, Version 090701, 2009 (accessed August 27, 2009). Available at
www.iau.org.tw/upload/download/1249000956.doc

[6] Kirk Reynolds, NCAA III Cross Country Team History - MEN (accessed June 10, 2010). Available at /www.pe.pomona.edu/sports/women/ wxc/ HistoriansReport/Men-team-history.pdf

[7] David E. Rosenbaum, The 2004 Campaign: the Independent—Relax, Nader advises alarmed Democrats, but the 2000 math counsels otherwise,*The New York Times*, February 24, 2004.

[8] Peter Tannenbaum, *Excursions in Modern Mathematics*, 6th ed., Pearson Education, Inc., Upper Saddle River, NJ, 2007.

[9] Alan D. Taylor, *Mathematics and Politics: Strategy, Voting, Power and Proof*, Springer-Verlag, New York, 1995.

About the Author

Steve Szydlik earned an M.A. in 1991 and a PhD. in mathematics in 1997 from the University of Wisconsin, and he has taught at the University of Wisconsin-Oshkosh since 1996. As an undergraduate at Union College, he ran both cross-country and track, and he remains a strong supporter of Division III athletics. His interest in cross country scoring systems arose when he was simultaneously teaching a course on the mathematics of voting and following the UW-Oshkosh cross country season.

CHAPTER 25

Biomechanics of Running and Walking

Anthony Tongen and Roshna E. Wunderlich

Abstract

Running and walking are integral to most sports and there is a considerable amount of mathematics involved in examining the forces produced when a foot contacts the ground. In this paper we discuss biomechanical terms related to running and walking. We use experimental ground reaction force data to calculate the impulse of running, speed-walking, and walking. We mathematically model the vertical ground reaction force curves for both running and walking, successfully reproducing experimental data. Finally, we discuss the biological implications of the mathematical models and give suggestions for classroom or research projects.

25.1 Introduction

Running speed is essential for many sports, whether it is the ability to beat a defender, run faster than an opponent, or develop enough take-off velocity to achieve distance or height on a jump. Running tends to occur at faster speeds than walking, although speed walkers can achieve speeds of up to 4.6 meters per second using an unusual gait in which the hip is dropped each step. Running is defined as a gait in which there is an aerial phase, a time when no limbs are touching the ground. Aside from wind resistance and gravity, there are no external forces applied to the body during this aerial phase. Therefore, it is the stance phase (the time when a limb is in contact with the ground) of running that must be modified in order to change speed.

We can measure the forces involved in the stance phase of running. This is called kinetics, which is the study of movement and the forces involved in producing it. Running forces are usually measured using a force plate. A force plate takes advantage of Newton's third law of motion: for every action there is an equal and opposite reaction. When we step on the ground we produce a vector of force that is generally downward and backward. The ground produces a force that is generally upward and forward, and it is this ground reaction force (GRF) that is measured by the force plate.

In order to run faster, stance time must be shorter. Shortening stance time, however, gives less time to produce an impulse, so the peak forces must be higher. Impulse is calculated by computing the impact force multiplied by the time over which it acts. The impulse for one step of a run is approximately constant regardless of the method of running. Consider the force on the knee joint for a stiff-legged running step and a compliant bent-knee step; recall that the force is the impulse divided by the impact time. For a stiff-legged run, the impact time is short, so the force on the knee is large. However, with a compliant bent-knee after impact, the impact time is longer, so the force on the knee is smaller (so bending their knees helps runners keep joint forces lower!). It is these high peak ground reaction forces at impact that contribute to the transmission of shock through the skeletal system and have been associated with overuse injuries such as shin splints and stress fractures [1, 2]. Mathematically, the impulse, I, of a step for running or walking is given by $I = \int F dt$, where F is the force.

Consider a three-dimensional coordinate system with orthogonal directions x, y, and z. Imagine walking along the x-axis, with the y-axis pointing toward the sky. Now think about the forces that the ground imparts on your every step as equal and opposite reactions; these are the forces measured by a force plate. The force with the largest magnitude that the ground imparts on your body is the vertical ground reaction force (VGRF), which is in the y-direction. The antero-posterior (fore-aft) force, in the x-direction, is less than the VGRF by approximately a factor of 10. The medial-lateral force, in the z-direction, is less than the VGRF by approximately a factor of 100 in both running and walking. Due to its importance, VGRF data will be the focus of this article. Data from one step of a run, walk, and speed-walk are shown in Figure 25.1.

It may be helpful to further examine the shape of the VGRF data in Figure 25.1 for each gait. The walking VGRF data exhibits two noticeable peaks. The first peak corresponds to the period just after the heel touches the force plate and the center of gravity is traveling down toward the ground, resulting in an increased reaction force from the ground in the vertical direction. The second peak corresponds to the toe pushing off of the force plate, applying a force into the ground which is matched by an increase in the ground reaction

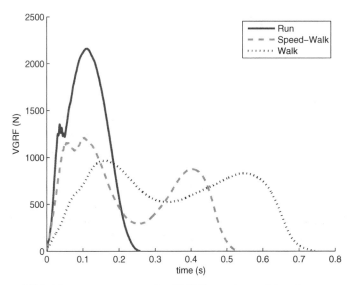

Figure 25.1. The curves correspond to VGRF experimental data for one step of a run (trial 8 from the spreadsheet), speed-walk (trial 1), and walk (trial 1).

force. The dip in the middle of these peaks occurs when the center of gravity is rising away from the ground, thus decreasing the force of the body on the ground and therefore decreasing the reaction force the ground is exerting on the body in the vertical direction. Two peaks are also observed in the speed-walking data, but there is an additional bump in the first peak called the impact peak. If you look back at the VGRF data for walking, you can also see a slight impact peak. The impact peak is the force associated with the foot striking the ground before the center of gravity travels downward, resulting in the larger VGRF peak. Finally, the running VGRF data has a single peak with a very sizable impact peak; in running the center of gravity is traveling down toward the ground on impact and changes directions only once to move away. As the center of gravity travels toward the plate, the body is increasing the force it is applying on the ground and therefore the ground is applying a greater force on the body.

Educationally, impulse is another useful application of integration. Since impulse is defined as the area under the curve of VGRF, $I = \int \text{VGRF} dt$, the data from the force plate can be used to illustrate the importance of either integration in general or numerical integration. Also, recall that the VGRF is a force; so using Newton's second law, $F = ma$ or $VGRF = ma$, we can find the acceleration to be $a = \frac{VGRF}{m}$. Now, in order to calculate the velocity, v, integrate the acceleration to obtain $v = \int \frac{\text{VGRF}}{m} dt$.

25.2 Applications

An Excel spreadsheet with data from walking, speed-walking, and running ground reaction forces is located at www.math.jmu.edu/~tongen/ TongenWunderlich.xls. These data were collected on a Kistler 9286AA force plate in the JMU Animal Movement laboratory. The subject weighed 77.51 kg (760.4 N), and the sampling rate of the force plate was 2500 Hz, which means that it calculated the forces every $\frac{1}{2500} = .0004$ seconds. In the spreadsheet, Fz is the VGRF, Fy is the antero-posterior force, and Fx is the medial-lateral force; all data have units of Newtons.

In the following, we use the spreadsheet data to calculate the impulse for each type of gait. We then use the data to model the VGRF curves for running while learning some biomechanical terminology. Finally, we use the data to model the VGRF curves for walking. After each discussion, we list possible classroom projects or research projects.

Numerically Calculating Impulse

Impulse describes the force applied over a period of time. It is given by $I = \int \text{VGRF} dt$. Looking at the data in Figure 25.1, which gait has the largest impulse? Does the height of the running curve overcome the width of the walking gait?

To calculate the impulse, we can approximate the area under the VGRF curve using Riemann sums. Recall from calculus that Riemann sums are usually introduced as a way of breaking a continuous curve into tiny rectangles and later trapezoids. Through Riemann sums, we calculate an approximation to the area under the curve. Many students cannot imagine a reason why one would ever want to make such an approximation. However, experimental data does not give a function describing the relationship between the independent and dependent variables. Therefore, calculating the impulse of the VGRF curve is a motivating example for numerical integration.

We will use the trapezoidal rule. Therefore,

$$I = \int \text{VGRF} dt \approx \sum_{k=1}^{n} \frac{F_k + F_{k+1}}{2} \cdot \Delta t,$$

where F_k is the force measurement from the force plate at time t_k and $\Delta t = t_k - t_{k-1}$, which is constant throughout the trial. Using the formula, we calculate the average impulse of the experimental data for running, speed-walking, and walking to be 309, 336, and 438 N·s, respectively.

While it is difficult to compare these values because we are only measuring one limb, and during walking and speed-walking the load of the body is shared with another limb, we can begin to examine the effects of speed using these data. We achieve a similar impulse in very different ways in running and speed walking. Running is characterized by high peak forces and short contact times, while walking is characterized by lower peak forces and longer contact times. At higher speeds, contact times will be even shorter, necessitating higher peak forces in order to support body weight.

Run with it

This section, and subsequent "Run with it" sections, will give ideas for classroom or research projects. Some of them will be accessible to students in courses as early as calculus, while others will require more advanced mathematical training.

1. Given a VGRF curve, find the velocity curve and interpret the results.

2. Given a VGRF curve, plot the position of the center of mass and interpret the results.

3. Verify the impulse-momentum relationship from the data.

Running Model

With a numerical approximation of the impulse of running in hand, we can now model the shape of the VGRF curve of a single step of a run. We initially assume that the VGRF curve is a parabola given by $F_R(t) = -at^2 + bt + c$, where $0 \leq t \leq T$ and T is the amount of time that the foot is in contact with the ground. The initial point of contact occurs at $t = 0$ while the last point of contact occurs at $t = T$. The first term of $F_R(t)$ is negative, because the VGRF curve is shaped like a parabola opening downward. Given $F_R(t)$, the impulse is given by $I = \frac{-aT^3}{3} + \frac{bT^2}{2} + cT$.

There are many ways to determine a, b, and c. Two logical boundary conditions are $F_R(0) = 0$ and $F_R(T) = 0$, which reduce the original equation to $F_R(t) = -a(t^2 - Tt) = -at(t - T)$ with only one unknown parameter a. Now we can use $F_R(t)$ to calculate the impulse, $I = \int_0^T F_R(t)dt = \frac{aT^3}{6}$. Therefore, the size of the impulse depends on the size of the yet unknown parameter a.

Visually, the most straightforward final assumption to find a, is that the maximum value occurs approximately at the midpoint of the VGRF curve, i.e., $F_R'(\frac{T}{2}) = 0$. This condition is automatically satisfied by our choice of a parabola to model the VGRF curve and the boundary conditions.

Since the maximum occurring at the midpoint is automatically satisfied, we can examine other possibilities to determine a. Assume the impulse, say I^*, is given or calculated from the experimental data. Then, we can determine the quadratic curve for VGRF in running as

$$F_R^{(1)}(t) = -\frac{6I^*}{T^3}t(t - T),$$

where I^* is the known impulse. We have called this approximation $F_R^{(1)}(t)$ to differentiate it from the following approximation.

Another possibility is to assume that the maximum VGRF value, M, is given, i.e. $F_R(T/2) = M$. Then, $F_R^{(2)}(t) = -\frac{4M}{T^2}t(t - T)$ with an impulse given by

$$I = \frac{2MT}{3}. \tag{25.1}$$

The size of the impulse depends on what happens to the value of MT, where M is the maximum value of the VGRF and T is the amount of time that the runner's foot is in contact with the ground. The mathematics completely agrees with the definition in the Introduction.

We can now use (25.1) to calculate the impulse of $F_R^{(2)}(t)$ from Figure 25.2. With $T = 0.2237$ seconds and $M = 2158.8$ N, the impulse is

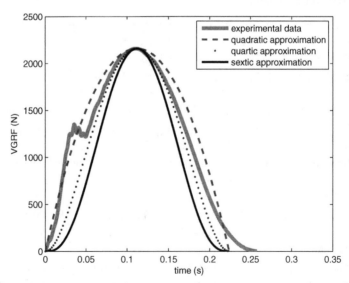

Figure 25.2. The thick curve corresponds to VGRF experimental data for one step of a run. The other curves are polynomial approximations to the VGRF curves with T=0.2237 s and M=2158.9 N.

$I = 321.964$ N·s, which is larger than the numerical value of 309 N·s. This discrepancy, along with the poor visual match of the quadratic approximation with the experimental data, motivates the following section where we introduce additional biomechanical concepts.

Inverse problem

Although the last section is a nice mathematical exercise, notice in Figure 25.2 that the quadratic approximation does a very poor job of approximating the VGRF curve for running. Before we increase the order of the approximating polynomial, we examine another biomechanical concept before returning to the mathematics.

Body weight, BW, can be approximated from the VGRF curve. A person running has to support an average of one body weight during the step cycle. Therefore, the BW is approximated by the line through the VGRF curve such that the area above the line but below the curve is equal to the area below the line and above the curve. Mathematically, the above concept for calculating BW is called the average value of a function, defined as

$$f_{\text{ave}} = \frac{1}{b - a} \int_a^b f(x)dx. \qquad (25.2)$$

We can calculate BW mathematically using the above definition to get

$$\int_0^{t_1} \left(BW - F_R(t)\right)dt + \int_{t_2}^T \left(BW - F_R(t)\right)dt$$
$$= \int_{t_1}^{t_2} \left(F_R(t) - BW\right)dt, \qquad (25.3)$$

where t_1 is the first intersection and t_2 is the second intersection of the line BW and $F_R(t)$. This simplifies to

$$BW \cdot t_1 - \int_0^{t_1} F_R(t)dt + BW(T - t_2) - \int_{t_2}^T F_R(t)dt$$
$$= \int_{t_1}^{t_2} F_R(t)dt - BW(t_2 - t_1), \qquad (25.4)$$

or $BW \cdot T = \int_0^T F_R(t)dt$, so $BW = \frac{\int_0^T F_R(t)dt}{T} = \frac{I}{T}$. By substituting $a = 0$ and $b = T$ into (25.2), it is straightforward to see that the calculation of BW is calculating an average value. Also, there is an immediate connection between body weight and impulse.

We can apply average value from the two quadratic approximations in the previous section. From the VGRF curve, where the maximum value M is easily calculated, using (25.1) we are able to find the body weight as $BW = \frac{2M}{3}$. Thus, knowing the BW allows us to approximate the maximum of the VGRF as $M = \frac{3}{2}BW$. However, our data show that the maximum VGRF values for a run are approximately 3 times BW, not $1\frac{1}{2}$ BW. Therefore, the curve does a poor job of approximating the VGRF curve as seen in Figure 25.2 and does not capture the biomechanics.

Suppose that the impulse I^* is known. As we saw earlier, one way to calculate impulse is to approximate it numerically from the experimental data. We can approximate $I^* = BW \cdot T$, body weight multiplied by the amount of time the foot is in contact with the ground. With this assumption, the maximum value can be determined by calculating $F_R^{(1)}(\frac{T}{2})$. After some algebra, we find $F_R^{(1)}(\frac{T}{2}) = \frac{3}{2}BW$. This result means that $F_R^{(1)} = F_R^{(2)}$ and both quadratic approximations to the VGRF curve in the previous section for running give the exact same answer, which is not a very good one.

Assuming a quartic approximation to the VGRF curve with boundary conditions given by $F_R(0) = 0$, $F_R(T) = 0$, $F_R'(0) = 0$, $F_R'(T) = 0$, and $F_R(T/2) = M$ gives

$$F_R(t) = \frac{-16M}{T^4}t^2(t - T)^2$$

and is the dotted curve in Figure 25.2. This implies that the maximum for the quartic approximation occurs at $F_R(\frac{T}{2}) = \frac{15}{8} \cdot BW$. Similarly, a sextic approximation to the VGRF curve with boundary conditions given by $F_R(0) = 0$, $F_R(T) = 0$, $F_R'(0) = 0$, $F_R'(T) = 0$, $F_R''(0) = 0$, $F_R''(T) = 0$, and $F_R(T/2) = M$ is given by

$$F_R(t) = \frac{64M}{T^6}t^3(t - T)^3$$

and is the thin solid curve in Figure 25.2. The maximum value occurs at

$$F_R(\frac{T}{2}) = \frac{35}{16} \cdot BW.$$

Although the sextic polynomial approximation is closer to predicting the maximum height of the curve based on the BW, visually it appears that the quartic approximation is the best. As we increase the order of the polynomial and continue to set higher derivatives at the boundary to zero, the polynomials will look more and more like a delta function and less and less like the experimental data. These results certainly motivate numerous projects for the "Run with it" section!

Run with it

Below are some classroom or research projects for students.

1. Are there other, more physically reasonable, ways to approximate the VGRF curve for running?

2. For the method and boundary conditions used, what is the limit of the maximum value as the order of approximation goes to infinity? Does the limit increase without bound or is it going to a limiting value?

3. We used polynomial interpolation, i.e., trying to model the data with a polynomial. Is it possible, and beneficial, to use Fourier series or another set of basis functions to model the data?

Walking Model

Figure 25.3 shows a typical VGRF curve for one step of a walk. This curve cannot be modeled by a quadratic function and a quartic function is the simplest first approximation. We will assume that the VGRF curve for walking is approximated by $F_W(t) = -at^4 + bt^3 + ct^2 + dt + f$, where $0 \leq t \leq T$. Modeling the VGRF curves for walking is more difficult than running, because of two additional unknown parameters. We will provide two examples of solving for the five unknown parameters, but we will leave it as an exercise to the reader to derive their own walking FUNctions!

Using the same boundary conditions that were used with running, we assume $F_W(0) = F_W(T) = 0$, which simplifies the approximation to

$$\begin{aligned}
F_W(t) &= -at(t^3 - T^3) + b(t^2 - T^2) + ct(t - T) \\
&= -t(t - T)(at^2 - bt + aTt - c - Tb + aT^2).
\end{aligned} \tag{25.5}$$

Next, assuming that $F_W'(\frac{T}{2}) = 0$, i.e., there is a local minimum in the middle, gives

$$\begin{aligned}
F_W(t) &= -at(t^3 - T^3) + 2aTt(t^2 - T^2) + ct(t - T) \\
&= -at(t^3 - 2Tt^2 + T^3) + ct(t - T) \\
&= -t(t - T)(at^2 - aTt - c - aT^2).
\end{aligned} \tag{25.6}$$

Two unknown parameters remain.

If the location of the maxima in the VGRF curve data is known, it will allow the determination of the two unknown parameters for the quartic function that best models the data. Assume that symmetric maxima occur at t_{max} and $T - t_{max}$, i.e., $F_W'(t_{max}) = F_W'(T - t_{max}) = 0$. These two conditions

yield only one unknown; as in running, they are duplicates and determine
only one of the parameters. We are left with

$$F_W(t) = -at(t^3 - 2Tt^2 + T^3) - a(T^2 - 2t \cdot t_{max} - 2t_{max}^2)t(t - T)$$
$$= -at(t - T)(t^2 - Tt - 2t_{max}^2 + 2T \cdot t_{max}). \tag{25.7}$$

With running we assumed that the impulse, I^*, was known, which led
to the same polynomial approximation as assuming that the maximum was
known. Therefore, for walking we will make assumptions about the function
$F_W(t)$ to avoid a similar redundancy. Assuming $F_W(t_{max}) = M$, we find

$$F_W^{(1)}(t) =$$
$$-\frac{t\left(T^2 t - 2t_{max}T^2 + 2Tt_{max}^2 - 2t^2T + 2Ttt_{max} + t^3 - 2tt_{max}^2\right)M}{(-t_{max} + T)^2 t_{max}^2},$$

which is shown by the dashed curve in Figure 25.3.

Another quartic approximation with $F_W^{(2)}(0) = 0$, $F_W^{(2)}(T) = 0$, $F_W'^{(2)}(\frac{T}{2})$
$= 0$, $F_W^{(2)}(t_{max}) = M$, and $F_W^{(2)}(\frac{T}{2}) = m$ is shown by the dotted curve
in Figure 25.3. The sextic polynomial approximation with $F_W(0) = 0$,
$F_W(T) = 0$, $F_W'(0) = 0$, $F_W'(T) = 0$, $F_W'(\frac{T}{2}) = 0$, $F_W(t_{max}) = M$,
and $F_W(T/2) = m$ is shown by the thin solid curve in Figure 25.3.

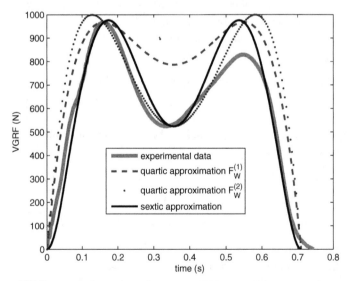

Figure 25.3. The thick curve corresponds to VGRF experimental data for one step
of a walk. The other curves are polynomial approximations to the VGRF curves with
$T = 0.7099$ s, $M = 966.86$ N, $t_{max} = 0.16$ s, and $m = 524$ N.

With walking, higher order polynomials visually do a better job of approximating the experimental data. Once again, we can calculate the impulse for the approximations to be 560.86 N·s for $F_W^{(1)}(t)$, 518.0 N·s for $F_W^{(2)}(t)$, and 460.9 N·s for $F_W(t)$. The calculated impulse from the experimental data was 438 N·s.

In running, we were able to use the idea of the average value of the function to assess the validity of the model. However, during a walk there is always a time when both feet are on the ground, i.e., there is no aerial phase. Therefore, the concept of calculating the average value doesn't apply to the walking gait.

Run with it

Below are some ideas for student projects.

1. Are there other, more physically reasonable, ways to approximate the VGRF curve for walking?

2. The modeling assumes that the two peaks in the VGRF curves are of equal height. Is there a way to incorporate a curve having two different peak heights into the model?

3. We used polynomial interpolation, i.e., trying to use a polynomial to model the data. Is it possible, and beneficial, to use Fourier series or another set of basis functions to model the data?

4. An interesting question is: what is happening in the transition between walking and running? We may conjecture that speed-walking is that transition, but why? In [3], a three-dimensional surface is made from experimental data to interpolate what occurs during the transition. It is straightforward to reproduce a modification of the surface [3], shown in Figure 25.4, given the models above. However, it doesn't make sense to have the impact time be the same in these different gaits. What modifications can be made to the surface to make it more physically useful? Can the surface answer any questions about speed-walking?

25.3 Conclusions

Sports biomechanics is replete with mathematical questions aimed at enhancing performance while simultaneously reducing the risk of injury. Injury usually occurs because of an overload to the musculoskeletal system. It is essential that we understand the forces involved in producing athletic

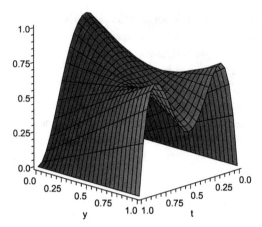

Figure 25.4. Surface to approximate the transition from walking to running. This surface uses the function $f(t, y) = F_R(t) \cdot (1 - y) + F_W(t) \cdot y$ where we used a quartic approximation for both $F_R(t)$ and $F_W(t)$.

movements as well as the features of these forces and movements that are associated with injury.

In this paper we have provided an introduction to some of the terminology and concepts of gait analysis using vertical ground reaction force in biomechanics. Mathematically we have only scratched the surface by using polynomials to model the VGRF curves that are seen in experimental data. We have also presented some applications of integration to the analysis of human gait, as well as some experimental data that can be used for educational purposes.

Acknowledgements   This work is supported by the NSF grant DMS-0734284. The authors thank Glenn Young for assistance with collecting and organizing data. The authors also thank D. Brian Walton for valuable discussions.

References

[1] C.E. Milner, I.S. Davis, and J. Hamill, Free moment as a predictor of tibial stress fracture in distance runners, *Journal of Biomechanics* 39 (2006) 2819–2825.

[2] C.E. Milner, J. Hamill, and I. Davis, Are knee mechanics during early stance related to tibial stress fracture in runners? *Clinical Biomechanics* 22 (2007) 697–703.

[3] T.S. Keller, A.M. Weisberger, J.L. Ray, S.S. Hasan, R.G. Shiavi, and D.M. Spengler, Relationship between vertical ground reaction force and speed during walking, slow jogging, and running, *Clinical Biomechanics* 11 (1996) 253–259.

About the Authors

Anthony Tongen is an Associate Professor in the Department of Mathematics and Statistics at James Madison University. In his free time, Anthony likes to play sports of all sorts as long as running is a means to an end.

Roshna E. Wunderlich is an Associate Professor in the Department of Biology at James Madison University. Roshna studies locomotion in humans and nonhuman primates and has the opportunity to watch biomechanics in action while coaching rugby.

About the Editor

Joseph A. Gallian was born in Arnold, Pennsylvania on January 5, 1942. He obtained a B.A. from Slippery Rock University in 1966, an M.A. from the University of Kansas in 1968 and a Ph.D. from the University of Notre Dame in 1971. After serving as a visiting Assistant Professor at Notre Dame for one year, he went to the University of Minnesota Duluth where he is a University Distinguished Professor of Teaching.

Among his honors are the MAA's Haimo Award for distinguished teaching, the MAA Allendoerfer and Evans awards for exposition, MAA Pólya Lecturer, MAA Second Vice President, MAA President, co-director of the MAA Project NExT, associate editor of the *American Mathematical Monthly* and the *Mathematics Magazine*, advisory board member for *Math Horizons*, the Carnegie Foundation for the Advancement of Teaching Minnesota Professor of the Year, and recipient of the University of Minnesota Duluth Chancellor's Award for Distinguished Research.

Over 150 research papers written under Gallian's supervision by undergraduates have been published in mainstream journals. He has given more than 250 invited lectures at conferences and colleges and universities and written more than 100 articles. He is the author of *Contemporary Abstract Algebra*, Cengage, 7th edition, coauthor of *For All Practical Purposes*, W.H. Freeman, 8th edition, and coauthor of *Principles and Practices of Mathematics*, Springer. He is the editor of two conference preceedings published by the American Mathematical Society and the Executive Producer of the documentary film "Hard Problems: The Road to the World's Toughest Math Contest." Gallian has received more than $4,000,000 in grants.

Besides the usual math courses, Gallian has taught a Humanities course called the "The Lives and Music of the Beatles" for more than 30 years and a liberal arts course on math and sports. In 2000 a Duluth newspaper cited him as one of the "100 Great Duluthians of the 20th Century."